Gha:

New Methods for Mining Sequential and Time Series Data

Ghazi Al-Naymat

New Methods for Mining Sequential and Time Series Data

Sequential and Time Series Data Mining

VDM Verlag Dr. Müller

Impressum/Imprint (nur für Deutschland/ only for Germany)

Bibliografische Information der Deutschen Nationalbibliothek: Die Deutsche Nationalbibliothek verzeichnet diese Publikation in der Deutschen Nationalbibliografie; detaillierte bibliografische Daten sind im Internet über http://dnb.d-nb.de abrufbar.

Alle in diesem Buch genannten Marken und Produktnamen unterliegen warenzeichen-, marken- oder patentrechtlichem Schutz bzw. sind Warenzeichen oder eingetragene Warenzeichen der jeweiligen Inhaber. Die Wiedergabe von Marken, Produktnamen, Gebrauchsnamen, Handelsnamen, Warenbezeichnungen u.s.w. in diesem Werk berechtigt auch ohne besondere Kennzeichnung nicht zu der Annahme, dass solche Namen im Sinne der Warenzeichen- und Markenschutzgesetzgebung als frei zu betrachten wären und daher von jedermann benutzt werden dürften.

Coverbild: www.purestockx.com

Verlag: VDM Verlag Dr. Müller Aktiengesellschaft & Co. KG
Dudweiler Landstr. 99, 66123 Saarbrücken, Deutschland
Telefon +49 681 9100-698, Telefax +49 681 9100-988, Email: info@vdm-verlag.de

Herstellung in Deutschland:
Schaltungsdienst Lange o.H.G., Berlin
Books on Demand GmbH, Norderstedt
Reha GmbH, Saarbrücken
Amazon Distribution GmbH, Leipzig
ISBN: 978-3-639-25774-8

Imprint (only for USA, GB)

Bibliographic information published by the Deutsche Nationalbibliothek: The Deutsche Nationalbibliothek lists this publication in the Deutsche Nationalbibliografie; detailed bibliographic data are available in the Internet at http://dnb.d-nb.de .

Any brand names and product names mentioned in this book are subject to trademark, brand or patent protection and are trademarks or registered trademarks of their respective holders. The use of brand names, product names, common names, trade names, product descriptions etc. even without a particular marking in this works is in no way to be construed to mean that such names may be regarded as unrestricted in respect of trademark and brand protection legislation and could thus be used by anyone.

Cover image: www.purestockx.com

Publisher:
VDM Verlag Dr. Müller Aktiengesellschaft & Co. KG
Dudweiler Landstr. 99, 66123 Saarbrücken, Germany
Phone +49 681 9100-698, Fax +49 681 9100-988, Email: info@vdm-publishing.com

Printed in the U.S.A.
Printed in the U.K. by (see last page)
ISBN: 978-3-639-25774-8

NEW METHODS FOR MINING SEQUENTIAL AND TIME SERIES DATA

by

Ghazi H. Al-Naymat

THE UNIVERSITY OF SYDNEY

May 2010

To my family

Whatever you do will be insignificant, but it is very important that you do it.

Mahatma Gandhi (1869 - 1948)

Acknowledgements

Thanks to my supervisor, Associate Professor Sanjay Chawla, for his strong support and invaluable guidance throughout my degree. I would also like to thank my associate supervisor, Professor Albert Zomaya who provided me with his insightful comments and advices.

I am very grateful to my parents and all my family members for their love, continuous support and the everlasting encouragement. I hope I will make them proud of my achievements, as I am always proud of them.

My gratitude and appreciation also goes to Joachim Gudmundsson, Mohamed Medhat Gaber and Javid Taheri for the helpful and constructive discussions.

Thanks to my colleagues and friends, Abed Kasiss, Noor Indah, Lorraine Ryan, Tony Souter, Mohammed Al-Khatib, Neda Zamani, Mohammed Al-kasasbeh, Elizabeth Wu, Bavani Arunasalam, Florian Verhein, Tara McIntosh and Paul Yoo for their help, love, friendship and encouragement.

I would also like to thank the technical support staff in our school for being helpful during my study years.

This work was financially supported by the Capital Markets Research Center (CMCRC).

Glossary of Terms

This glossary contains the key concepts and terms that are used in this book. These concepts are defined in detail where they appear in the book chapters; however, they have been listed here for ease of reference.

Term	Description
Association rule	An implication expression of the form $A \to B$, where A and B are disjoint items, i.e., an expression that provides the relationship between the items A and B (Tan et al., 2006)
Clique	A group of objects such that all objects in that group are co-located with each other
Co-location pattern	A group of spatial objects so that each is located in the neighborhood of another object in the group
Co-location rule	A representation that signifies the presence or absence of spatial features in the neighborhood of other spatial objects (Huang et al., 2004)
Curse of dimensionality	A term to describe the problem caused by the exponential raise in volume associated with adding more dimensions (features) to a space (Bellman, 1961)

Term	Description
Data Mining	Is the process of analyzing large datasets from different perspectives and summarizing them into useful information. It is a synonym of the Knowledge Discovery in Database
Flock pattern	A group of objects that move close to each other for a period of time
Knowledge Discovery in Database	A process of extracting interesting (non-trivial, hidden, formerly unknown and potentially handy) information or patterns from large information repositories, such as relational database, data warehouses, XML repository
Long-buy	A practice of buying stocks in the hope of selling them in the future at a higher price
Maximal clique	A clique that does not appear as subset of another clique in the same co-location pattern
Pairs trading	An investment strategy that involves buying the undervalued security and short-selling the overvalued one, thus maintaining market neutrality
Random projections	A technique that tackles the curse of dimensionality, using sparse random matrices
Short-sell	A practice of selling stocks which are not owned by the seller, in the hope of repurchasing them in the future at a lower price
Sparse matrix	A matrix populated mostly with zeros

Term	Description
Spatial Data	Data or information that identifies the geographic location of features and boundaries on Earth, such as natural or constructed features, oceans, and more. Spatial data is usually stored as coordinates and topology, and is data that can be mapped. It is known as geospatial data or geographic information
Spatio-Temporal Data	Data that manages both space and time information, such as biological Databases, wireless communication networks, and processing of objects with uncertainty
Spread	The difference between bid and ask prices
Time Series Data	A sequence of data points, measured typically at successive times, where each data points represents a value

Acronyms

Acronym	Expanded term
DBMS	Database Management System
TDBMS	Temporal Database Management System
DM	Data Mining
KDD	Knowledge Discovery in Database
S-Data	Spatial data
SIS	Spatial Information System
AM/FM	Automated Mapping Facilities Management
GIS	Geographic Information Systems
GPS	Global Positioning System
ST	Spatio-Temporal
TSD	Time Series Data
TSDM	Time Series Data Mining
DTW	Dynamic Time Warping
LCSS	Longest Common Subsequence
DC	Divide and Conquer
BandDTW	Sakoe-Chiba Band
SparseDTW	Sparse Dynamic Time Warping
EucDist	Euclidean Distance
$\mathbb{R}$	Real numbers
$\mathbb{N}$	Natural numbers

Acronym	Expanded term
DFT	Discrete Fourier Transform
DWT	Discrete Wavelet Transform
SVD	Single Value Decomposition
PCA	Principle Components Analysis
RP	Random Projections
REMO	RElative MOtion
$X_{n \times d}$	Original data set
$R_{n \times \kappa}$	Random matrix
$rand(\cdot)$	Function to generate random numbers between 0 and 1
$princomp(\cdot)$	Function to retrieve the principle Components of a matrix
$MaxPI$	Maximum Participation Index
$\lvert S \rvert$	The length of the time series S
dist	Distance function
ϵ	A real value for a predefined matching threshold
$\mathcal{D}$	A set of time series (sequences) data
D	Warping Matrix
S	Time series (Sequence)
Q	Query time series (Sequence)
C	Candidate time series (Sequence)
s_i	The ith element of sequence S
q_i	The ith element of sequence Q
P_i	The price of stock i
nP_i	Normalized price of stock i
$DTW(Q,S)$	DTW distance between two time series Q and S
$LB\_Keogh(Q,C)$	Lower bound for query sequence Q and candidate sequence C
SP	Stock pair
k-NN	k Nearest Neighbors
MBR	Minimum Bounding Rectangle

Acronym	Expanded term
d	Number of dimensions
n, m	Number of data points such as rows, objects
r	Radius
τ	Number of time steps
κ	Number of desired dimensions
MCCR	Mining Complex Co-location Rules
parsec	Unit of length used in astronomy. It stands for "**par**allax of one arc **sec**ond"
Mpc	An abbreviation of "Mega-parsec", which is one million parsecs, or 3261564 light years
arcmin	Unit of angular measurement. Sizes of objects on the sky, field of view of telescopes, or practically any angular distance "Arc of Minutes"
z	RedShift
zWarning	Parameter used to guarantee that the corrected RedShift values are used
zConf	RedShift confidence
U	Ultraviolet
R	Red light magnitude
r-band	r-band Petrosian magnitude
H_o	Hubble's constant
LRG	Luminous Red Galaxies
minPI	Minimum Participation Index
maxPI	Maximum Participation Index
SDSS	Sloan Digital Sky Survey
O	Spatial objects with fixed locations
o_i	The ith spatial object
G	Undirected graph
C_M	Maximal Clique
T	Set of transactions

Acronym	Expanded term
$g(\cdot)$	Function that performs a transformation on the transposed dataset
$\circ$	An operator that combined two *itemvectors* together to create a new *itemvector* corresponding to the union of the two itemsets
P	A set of complex types
$N(\cdot)$	Number of maximal cliques that contain the set of complex types
I	Item set
$f(\cdot)$	Item set size
$|\cdot|$	Number of elements in a set
$Card(\cdot)$	Cardinality of a set
X	X coordinate
Y	Y coordinate
Z	Z-coordinate
w/o	Without
w	With
$\lceil\cdot\rceil$	Function that gives the ceiling of a value
$\lfloor\cdot\rfloor$	Function that gives the floor of a value
NDSSL	Network Dynamics and Simulation Science Laboratory
R	Random matrix
r_{ij}	An entry of a matrix
ASX	Australian Stock eXchange
LBF	Lower Bound Function
SM	Sparse Matrix
W	Warping path
SIRCA	Securities Industry Research Centre of Asia-Pacific
UCR	University of California - Riverside

Contents

List of Tables

List of Figures

Chapter 1

Introduction

Rapid advances in data collection technology have enabled organizations and businesses to store massive amounts of data. This growth in the size of datasets has meant that it is now beyond human capacity to analyze them to discover useful information rules, hidden clusters and implicit regularities. A major problem has arisen, because it is hard to use traditional data analysis tools to analyze large datasets. Differences in the information stored by non-traditional datasets, which include spatial, spatio-temporal and time series, mean that traditional analysis tools cannot be used to analyze non-traditional data. Thus, new tools need to be designed and developed to mine the large collection of non-traditional datasets.

Data mining combines traditional data analysis methods with sophisticated techniques to process large volumes of data. Data mining includes a range of different techniques that reveal diverse kinds of patterns from a given database, based on the requirements of the application area. These techniques include association rules mining, classification, cluster analysis and outlier detection. Due to the availability of applications that produce large amounts of spatial, spatio-temporal (ST) and time series data (TSD), we have proposed in this research specialized data mining techniques to mine such data.

This book is composed of four parts. The first shows the need to develop efficient methods to mine complex patterns from large spatial datasets. The knowledge uncovered using these methods could help scientists to prove existing facts. The second pertains to mining long-duration and complex spatio-temporal patterns, such as flock patterns; it deals specifically with the problem of monitoring moving objects, with potential significance to organizational planning and decision-making processes. The third part introduces an efficient algorithm to the area of time series mining, which can be used to mine similarity between time series. The last part shows the successful application of our algorithm (proposed in the third part) to successfully and correctly discover interesting patterns (pairs trading) from real-world data.

The chapter is organized as follows. Section 1.1 gives an overview of research problems, and provides the rationale behind the approaches used in the project. Section 1.2 summarizes the objectives and contributions of the book, followed by its organization in Section ??.

1.1 Background and Rationale

This section provides a motivation for the book by discussing a number of open problems.

1.1.1 Mining Complex Co-location Rules (MCCRs)

Spatial databases regularly contain not only traditional data, but the location or geographic details of the corresponding data; in other words, spatial data captures information about our surroundings by storing both conventional (*aspatial*) data and *spatial* data. Spatial data is often described as *geo-spatial* data. In this book, we use a large astronomy dataset containing the location of different *types* of galaxies. The widespread use of spatial databases is leading to a rising concentration in the mining of interesting and valuable but implicit patterns (Koperski et al., 1996). One such pattern is the *co-location* pattern – group of objects (such as galaxies) located so that each is in the neighborhood (within a given distance) of another object in the group.

Mining spatial co-location patterns is a significant spatial data-mining task with wide-ranging applications (Huang et al., 2004); examples include public health, environmental management, transportation and ecology. The mining of co-location patterns is very important, since its results can be used to mine interesting co-location rules; i.e., co-location patterns will be represented as the raw data for the mining rules process.

In this book, we will mine complex co-location rules, which are defined as representations that indicate the presence or absence of spatial features in the neighborhood of other spatial objects (Huang et al., 2004). The complex rules are a combination of two different types of rules. The first, *positive rules*, defines the presence of one or more objects of the same type in the neighborhood of another object. This rule appears as $A \rightarrow B$ or $A+ \rightarrow B$, where the sign $(+)$ indicates the presence of more than one object-type. The second, *negative rules*, defines the absence of an object from the neighborhood of another object – for example, $-A \rightarrow B$, where the sign $(-)$ indicates the absence of an object. To give a real-life example of such rules, we will provide two rules from the astronomy domain, such as {elliptical galaxy} $\rightarrow$ {spiral galaxy} and {elliptical galaxy} $\rightarrow$ {−spiral galaxy}. The last is interpreted as the presence of an elliptical galaxy, implying the absence of a spiral galaxy.

The transportation development domain is another interesting example that shows that mining complex rules provides more detailed insights into the domain. The rule {more trains} $\rightarrow$ {quick service} might be a valid positive rule, and the rule {− maintenance} $\rightarrow$ {slow service} might be a valid negative rule. The question is: what is the implication of the combination {more trains, − maintenance}? Positive or negative rules by them-selves will be unable to provide us with knowledge that may found from the presence or absence of services and facilities together. However, a complex rule might indicate that {more trains; − maintenance}→ {bad service}. This rule offers more insight into the transportation development management than the other.

To mine complex co-location rules, two major points need to be considered. The first is that the spatial data must be transformed into a transactional-type dataset – to allow the association rule mining technique to be applied. The transformation is performed by extracting co-location patterns such as clique patterns from the raw spatial data. A clique is a special type of co-location pattern, which can be described as a group of objects in which *all* objects are co-located with each other. The second point is discovering *maximal cliques*, which are cliques that do not appear as a subset of another clique in the same co-location pattern. The problem of extracting maximal clique patterns is NP-hard problem (Arora and Lund, 1997). Mining maximal clique patterns allows us to mine interesting *complex spatial relationships* between the object types, as will be described in Chapter 3.

Huang et al. (2004) defined co-location patterns as the presence of a spatial feature in the neighborhood of instances of other spatial features. The authors developed an algorithm for mining valid rules in spatial databases, using an Apriori-based approach. Their algorithm does not separate the co-location mining and interesting pattern-mining steps. The authors did not consider complex relationships or patterns, because they were pruning most items on the basis of their prevalence measure, known as the "maximum participation index" (MaxPI); however, these items might contribute to forming complex rules. Munro et al. (2003) used cliques as a co-location pattern, and in an approach similar to ours, they separated the clique mining from the pattern-mining stages; however, they did not use maximal cliques. Arunasalam et al. (2005) used a similar approach to Munro et al. (2003), proposing an algorithm (NP_maxPI) that also used the MaxPI measure.

The main motivations behind our proposed approach in Chapter 3 are:

1. Previous approaches used the concept of clique to mine complex co-location patterns. This allows redundancy in the co-location pattern itself as well as precluding the inference of the negative relationship between objects. However, the use of maximal clique patterns makes more sense in the mining of complex co-location patterns, because it ensures that all of the members are co-located; this means that it is possible to infer negative relationship (relationships which indicate the absence of some items that gives useful information about other present items). Therefore, this requires the development of an efficient algorithm to extract maximal clique patterns.

2. All previous approaches used Apriori-type algorithms, which are not efficient for mining large datasets consisting of complex relationships; this has motivated us to use an efficient algorithm, called GLIMIT.

1.1.2 Mining Complex Spatio-temporal Patterns

Spatio-temporal (ST) data contains the evolution of objects over time as well as their spatial features (Tsoukatos and Gunopulos, 2001). A wide range of scientific and business applications need to capture the ST characteristics of the entities (objects) they model. ST applications are becoming popular with the increasing abilities of computer systems to store and process large amounts of data. Examples of such applications include land management, weather monitoring, natural resources management and the tracking of mobile devices. Another reason for the availability of ST data is the widespread use of GPS-enabled mobile devices and location-aware sensors. A distinctive example of a project that is continuously producing ST data is related to the tracking of caribou in Northern Canada. Since 1993, the movement of caribou has been tracked through the use of GPS collars, with the underlying hope that the data collected will help scientists to understand the migration patterns of caribou and to locate their breeding and calving locations (pch, 2007). While the number of caribou tagged at a given time is small, the time interval (the temporal data) for each animal is long.

In data mining research, the focus is to design techniques for discovering new patterns in large repositories of ST data – for example, Mamoulis et al. (2004) mine periodic patterns moving between objects. More recently, Verhein and Chawla (2008) have proposed efficient association mining-type algorithms to discover ST patterns such as sinks, sources, stationary regions and thoroughfares. In this book, we focus on the fixed-subset flock pattern (where objects are moving close together in coordination). Benkert et al. (2006) described efficient approximation algorithms for reporting and detecting flocks. Their main approach is a $(2 + \varepsilon)$-approximation, where the dependency on the duration of the flock pattern is exponential.

Mining moving object patterns is an important problem in many domains. This book focuses on mining "flock-query" – that is, a query that reports group of moving objects that move together for a period of time. This query can readily be applied to better understand the scenarios below:

1. The "pandemic preparedness" studies, which have an ultimate goals to answer number of question, such as "How does a contagious disease get transmitted in a large city given the movement of people across the city?"

2. Applications in the area of defence and surveillance, where analysts aim to obtain knowledge about patterns that might be of interest, such as smugglers or terrorist groups.

Previous approaches have developed algorithm to report flock patterns; however, they were unable to report long-duration patterns. The reason was the exponential dependency on the pattern duration (the trajectory length). This requires the development of a robust algorithm with a smaller dependency on duration (number of dimensions in the ST data). This has become the reason for our proposed approach in Chapter 4.

1.1.3 Mining Large Time Series Data

A time series is a sequence of data points that are typically measured at successive time intervals, where each data point represents a value; therefore, time series data (TSD) is a collection of sequences. In the remainder of the book, the terms "time series" and "sequence" are used interchangeably.

Similarity searches in TSD is very popular (Sakurai et al., 2005). The similarity can be evaluated using distance measures, such as Euclidean distance or dynamic time warping (DTW). Since TSD normally consists of sequences of different length as well as out of phase, DTW is a highly accepted mechanism because it allows sequences to be stretched along the time axis to minimize the distance between the sequences. Chapter 2 uses an example to highlight the differences between Euclidean distance and DTW.

DTW uses the dynamic programming paradigm to compute the alignment between two time series. An *alignment* "warps" one time series onto another, and can be used as a basis to determine the similarity between the time series. The standard *DTW* algorithm has $O(mn)$ space complexity, where m and n are the lengths of the two sequences being aligned.

Given the expensive space complexity of DTW, many researchers have developed techniques to increase the speed of DTW. A brief categorization of these techniques includes those that add constraints on DTW to reduce the search space, such as the work performed by Sakoe and Chiba (1978) and Itakura (1975). While these approaches provide a reduction in space complexity, they do not guarantee the optimality of the alignment. The second technique is based on data abstraction, where the warping path is computed at a lower resolution of the data and then mapped back to the original resolution (Salvador and Chan, 2007). Again, discovering the optimal alignment is not guaranteed. The third technique, indexing techniques, such as those proposed by Keogh and Ratanamahatana (2004) and Sakurai et al. (2005) that do not directly speed up *DTW*, but limit the number of *DTW* computations.

TSD is naturally produced by many different applications, such as computational biology and economics. The data generated by those applications continues to grow in size, placing increased demand on developing tools that capture the similarities among them. Interesting examples of real-life queries that can be answered by reporting the similarity between sequences are:

1. Financial sequence matching, where investors intend to monitor the movement of stock prices to obtain information about price-changing patterns or stocks that have similar movement patterns. One of the most sought-after patterns in this sector is called "pairs trading", where investors seek knowledge to make more profit and reduce their expected loss. Our approach will be applied in large stock data to mine such patterns, which will be described in Chapter 6.

2. Speech recognition applications that handle large audio/voice data. For example, analysts aim to answer a query such as "find clips that sound like a given person".

Most researchers have tried to increase the speed of DTW as the underlying similarity measure in time series mining, by either reducing the space search with the sacrifice of some accuracy, or proposing lower bounding techniques, which reduce the number of

DTW computations rather than the computational time. Because of these limitations, we have devised an algorithm that reduces the DTW search space while guaranteeing accuracy as will be described in Chapter 5.

1.1.4 Mining Pairs Trading Patterns

Many researchers have developed algorithms and frameworks that concentrate on mining useful patterns in stock market datasets. The literature demonstrates that *pairs trading* is one of the most sought-after patterns because of its market-neutral strategy (Vidya-murthy, 2004). Pairs trading is an investment strategy that involves buying undervalued stock, while short-selling the overvalued, thus maintaining market neutrality. Finding pairs trading is one of the pivotal issues in the stock market, because investors tend to conceal from others their prior knowledge about the stocks that form pairs, to gain the greatest advantage.

Several methods have been proposed to mine pairs trading. Association rules are used to predict the movement of the stock prices, based on recorded data (Lu et al., 1998; Ellatif, 2007); this will help to find the convergence in stock prices. However, association rule mining techniques usually generate a large number of rules, which presents a major interpretation challenge for investors. Basalto et al. (2004) have applied a non-parametric clustering method to search for correlations between stocks. Cao et al. (2006b) introduced fuzzy genetic algorithms to mine pair relationships and proposed strategies for the fuzzy aggregation and ranking to generate the optimal pairs for the decision-making process.

Investors who monitor stock price movements (changes) are often looking for patterns that are indicative of profit. It would be a great advantage to use a computer to monitor the evolution of stock prices, to assist investors to make optimal decisions when buying and selling stocks. This should be achieved by reporting all stock pairs (stocks that are most similar in their price-movement profiles). This is the main motivation for our approach in Chapter 6, where we efficiently mine pairs trading patterns in large stock market data, using DTW as the underlying similarity measure. To the best of our knowledge, previous approaches have never used DTW for the purpose of mining pairs trading patterns. Since DTW has been used as a successful shape-similarity measure, we have used it to monitor time series similarity.

1.2   Objectives and Contributions

To summarize, the main objectives and contributions are given in the following subsections.

1.2.1   Mining Complex Co-location Rules

This section sets out our main objectives and contributions to the area of complex co-location rules mining. The ultimate goal is to efficiently mine complex co-location rules from a large spatial dataset (astronomy dataset). To accomplish this, we propose an algorithm (GridClique) based on a divide-and-conquer strategy, to efficiently mine maximal clique patterns. We show that in conjunction with the GridClique algorithm, any association rule mining technique can be used to mine complex, interesting co-location rules efficiently. The results from our experiments, which are carried out on a real-life dataset obtained from an astronomical source – the Sloan Digital Sky Survey (SDSS) – are of potentially valuable to the field of astronomy and can be interpreted and compared easily to existing knowledge.

1.2.2 Mining Complex Spatio-Temporal Patterns

We summarize our objectives and contributions to the area of mining complex spatio-temporal patterns (flock patterns). Our main objective is to efficiently report long-duration flock patterns in large ST datasets. To achieve this, we use a new approach that combines random projections, as a dimensionality reduction technique, with an approximation algorithm. To the best of our knowledge, this is the first time that random projection has been used to reduce dimensionality in a the ST setting presented in this book. We prove that the random projection will return the "correct" answer with high probability. Our experiments on real, quasi-synthetic and synthetic datasets strongly support our theoretical bounds.

1.2.3 Mining Large Time Series Data

In this section, we summerize our objective and contribution to the area of mining large time series data.

The main objective is to speed up the computation of DTW as the similarity measure without sacrificing any accuracy. To attain this, we devise an efficient algorithm (SparseDTW) that exploits the possible existence of inherent similarity and correlation between the two time series whose DTW is being computed. We always represent the warping matrix using sparse matrices, which lead to better average space complexity compared with other approaches. The SparseDTW technique can easily be used in conjunction with lower bounding approaches. Our experiments show that SparseDTW gives exact results, unlike other techniques, which give approximate or non-optimal results.

1.2.4 Mining Pairs Trading Patterns

A summary of our objectives and contribution, to the area of mining financial data, is described in this section. To help investors in the finance sector to make profit and reduce the risk of their investments, our goal is to report to those investors, accurately, all pairs patterns from large daily TSD (e.g., stock market data). To accomplish this, we propose a framework to successfully find pairs trading patterns in large stock market data, using DTW as a the similarity between stocks and by applying our algorithm SparseDTW to reports all pairs. Our experiments show that SparseDTW is a robust tool for mining pairs trading patterns in large TSD.

1.3 Organization of the Book

The book is structured as follows:

Chapter 2 introduces the key concepts and the foundations of three different areas. These are spatial, ST and time series data mining. It provides a review of the recent previous work that has been conducted in these three areas. In Chapter 3, we discuss the implementation of the proposed approach, that is Mining Complex Co-location Rules (MCCR). This work has been published in various conference proceedings (Al-Naymat, 2008; Verhein and Al-Naymat, 2007). Chapter 4 presents the dimensionality reduction approach (random projections) that has been used to mine long duration flock patterns. The experiments in this chapter show the correctness of the approach. This work has been published in technical reports and various conference proceedings (Al-Naymat et al., 2006, 2007, 2008a). In Chapter 5, we present our novel algorithm (SparseDTW), which used to mine similarity in large time series datasets. This work will appear in journal and (Al-Naymat et al., 2008b). Chapter 6 exhibits an interesting case study where SparseDTW is applied to successfully mine pairs trading patterns in large time series dataset (stock market dataset). This appears in (Al-Naymat et al., 2008b). A summary of the research conducted in this book, conclusion and future directions are presented in chapter 7.

There are three appendices. Appendix A demonstrates a comprehensive explanation for the data preparation stage that was performed on the large spatial dataset used in Chapter 3. In Appendix B, we present a description of the data preparation conducted on the ST datasets used in Chapter 4. Appendix C provides a detailed description of the process used to obtain the TSDs used in Chapters 5 and 6.

Chapter 2

Related Work

This chapter defines key concepts in the field of data mining, and presents an overview of previous work in the areas of spatial, spatio-temporal and time series mining. After the data mining and database overviews (Sections 2.1 and 2.2), the chapter reviews three main areas of the literature: spatial data mining (Section 2.3), spatio-temporal data mining and dimensionality reduction (Sections 2.4 and 2.5, respectively), and time series mining and the similarity measures used in the mining of large time series datasets (Section 2.6).

This chapter lays the foundations for the book. The research that is more specifically related to each of our proposed methods will be discussed in more detail in the Chapters 3, 4, 5 and 6. Table 2.1 lists the notations used in this chapter.

Symbol	Description
DM	Data Mining
KDD	Knowledge Discovery in Database
DBMS	Database Management System
TDBMS	Temporal Database Management System
S-Data	Spatial data.
GIS	Geographic Information Systems
GPS	Global Positioning System
ST	Spatio-Temporal
TSD	Time Series Data
TSDM	Time Series Data Mining
DTW	Dynamic Time Warping
LCSS	Longest Common Subsequence
DC	Divide and Conquer
BandDTW	Band constraint on DTW
SparseDTW	Sparse Dynamic Time Warping
EucDist	Euclidean distance
$\mathbb{R}$	Real numbers
$\mathbb{N}$	Natural numbers
DFT	Discrete Fourier Transform
DWT	Discrete Wavelet Transform
SVD	Single Value Decomposition
PCA	Principle Components Analysis
RP	Random Projections
REMO	RElative MOtion
MaxPI	Maximum Participation Index
$|S|$	The length of the time series S
S	Time series (Sequence)
Q	Query time series (Sequence)
C	Candidate time series (Sequence)
s_i	The ith element of sequence S
q_i	The ith element of sequence Q
dist	Distance function
$DTW(Q, S)$	DTW distance between two time series Q and S
ϵ	A real value for a predefined matching threshold
$\mathcal{D}$	A set of time series (sequences) data
D	Warping Matrix
k-NN	k Nearest Neighbors
MBR	Minimum Bounding Rectangle
d	Number of dimensions
n	Number of data points, such as rows, objects
r	Radius
τ	Number of time steps
κ	Number of desired dimensions
minPI	Minimum Participation Index

Table 2.1: Description of the notations used.

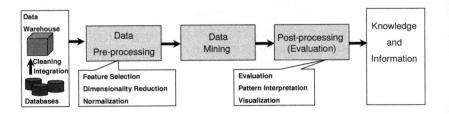

Figure 2.1: The process of Knowledge Discovery in Databases(KDD).

2.1   Data Mining Overview

Data mining is defined by Piatetsky-Shapiro and Frawley (1991) as the process of extracting interesting (non-trivial, hidden, formerly unknown and potentially useful) information or patterns from large information repositories such as relational databases, data
warehouses and XML repositories.

Although data mining is one of the core tasks of Knowledge Discovery in Databases
(KDD), many researchers understand it as a synonym for KDD (Figure 2.1). The KDD
process typically consists of four stages. The first is the pre-processing stage, which is
executed as a preliminary step before applying data mining. The pre-processing step
includes data cleansing, integration, selection and transformation. The main process of
KDD is data mining, in which different algorithms are applied to produce knowledge
that is out of sight. Following this, another process called pattern evaluation evaluates
all results according to users' requirements and domain knowledge. Once complete, an
evaluation displays the results if they are suitable; otherwise some or all of the KDD
stages have to be run again until suitable results are obtained. In greater detail, the KDD
stages work in the following sequence:

In the first step, it is mandatory to clean and integrate the databases. Since the data source

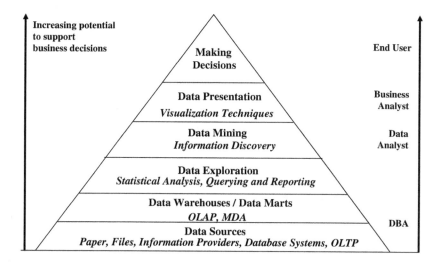

Figure 2.2: Data Mining and Business Intelligence (Han and Kamber, 2006).

may come from domains that may have inconsistencies and duplications, such as noisy data. The cleaning process is performed by removing such undesirable information. For example, if there are two different database resources, different attributes are used to refer to the same description in the schema. When integrating these two resources, one of these attributes can be chosen, and the other discarded. Real-world data tends to be incomplete and noisy due to manual input errors. One way of integrating these data sources is to store the data in a data warehouse, in the process removing redundant and noisy data.

Although databases are a good way to avoid redundancy and eliminate noise, the raw data is not necessarily used for data mining. Therefore, the second stage is to select the related data from the integrated resources and convert them into a format that is acceptable for data mining. For example, an end user who wants to find which items are often purchased together in a supermarket may find that the database recording the purchase history contains attributes such as *customer ID, items bought, transaction time, prices, quantity*. However, for this specific task, the only information necessary is a list of the purchased items. Consequently, selecting only the relevant information will reduce the size of the experimental database; this will have a positive impact on the efficiency of the KDD process.

After pre-processing the data, a variety of data mining techniques can be applied. Different data mining techniques allow the discovery of different knowledge, which needs to be evaluated according to certain rules, such as domain knowledge or user requirements. The last stage in the KDD process is the evaluation (Figure 2.1). In this stage, the produced results are matched with the user's requirements or the domain rules. If the results do not suit the domain or the end user's requirements, two procedures may be applied. First, the mining process must be run until the desired results are achieved and/or the requirements must be modified. One of the main steps of the stage of evaluating the results is to visualize them. This helps users to understand and interpret the results in a way that is meaningful to them and meets their desired purpose. The results can be visualized in tools such as tables, decision trees, rules, charts or 3D graphics. Visualization is normally achieved by the *business analyst* as shown in Figure 2.2. Ultimately, the end user (top of the pyramid in Figure 2.2), will use the produced knowledge in the decision-making process.

An example is the market basket applications that produce daily massive amounts of transactional data. If we apply the association rule mining technique, the produced knowledge (rules) will be of the form $\{antecedent \rightarrow consequent\}$ – for example, $\{bread \rightarrow cheese\}$. This signifies that customers who buy bread are likely to buy cheese as well. Such rules are useful for product pricing, promotion and placement, and when making decisions on store management and organization.

2.2 Overview of Databases

After the overview on the KDD process, this section will provide a general overview on databases relevant to this research.

Saraee and Theodoulidis (1995) categorized databases into four groups, or shapes: (1) *snapshot databases* (conventional databases without time "past state"); (2) *rollback databases* (those that store each transaction's creation time); (3) *historical databases* (those that store the real time for each event); and (4) *bi-temporal databases*, which merge rollback and historical databases (which together are also known as *temporal databases*). In other words, bi-temporal databases contain both valid and transaction time-stamp for the event. Lopez and Moon (2005) defined *valid time* as the time at issue when the event is true; they defined *transaction time* as the time at issue when the event is in the database.

The focus of this book is to develop new methods for mining a mixture of these four types of databases. Specifically, this will involve *spatial data*, which adds a location feature to the conventional databases; *spatio-temporal data*, which is spatial data that captures the time-stamps for each object location; and *time series data*, which is defined as a sequence of data points, typically measured at successive periods/intervals.

The following sections review the literature on *spatial*, *spatio-temporal*, and *time series* data mining.

2.3 Spatial Data

Spatial databases usually contain not only traditional data, but the location or geographic information about the corresponding data. In other words, spatial data captures information about our surroundings by storing both conventional (*aspatial*) data and *spatial* data. Judd (2005) defined spatial data as location-based data. However, a spatial database

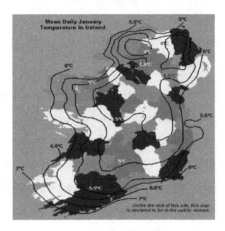

(a) Example of temperature patterns. Mean daily temperature in Ireland over January (Ire, 2008).

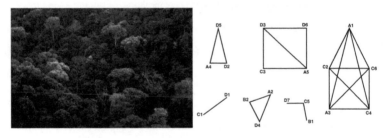

(b) Example of landscape patterns.   (c) Example of co-location patterns.

Figure 2.3: Examples of spatial patterns.

is defined simply as a collection of data that contains information on observable facts of interest, such as daily temperature over a period of time (Figure 2.3(a)), forest condition (Figure 2.3(b)), and the location of observable objects in any space (Figure 2.3(c)).

Spatial databases and their related processes are handled by a system named Spatial Information System (SIS). This system is categorized into four systems: Geographical Information Systems (GIS), Automated Mapping Facilities Management (AM/FM) Systems, Land Information Systems (LIS), and Image Processing Systems (Egenhofer, 1993). According to Güting (1994), there are three different architecture types of SIS: dual, layered, and integrated architectures.

Jeremy and Wei (2005) mentioned several tasks for GIS, such as integrating diverse datasets, extracting spatial relationships and classifying numeric data into ordinal categories (useful when applying the association rule mining technique).

Spatial data has different granularities, which show the level of detail of the examined space, and thus define the characteristics of the spatial domain (Tsoukatos and Gunopulos, 2001).

2.3.1   Spatial Patterns Mining

Although there are many types of spatial patterns, the research will focus on the discovery of the co-location patterns. These patterns are defined on the basis of the relationships between the spatial objects. For example, the neighborhood relationships between spatial objects indicate that objects are co-located. To define a co-location relationship, there should be at least two spatial objects which are close to each other (neighbors). Two objects are neighbors if the distance between them is less than or equal to a predefined distance.

Co-location patterns are considered one of the most complex and interesting spatial relationships that can be reported from large spatial databases. Figure 2.3(c) illustrates an example of spatial objects forming co-location patterns. The set of spatial objects $\{A2, B2, D4\}$ is an example of a co-location pattern. This section provides an overview of the techniques used to retrieve different types of co-location patterns.

Morimoto (2001) defined new co-location patterns that he called the $k - neighboring$ class-set. His method uses the number of instances of a pattern as a prevalence measure. However, this measure does not satisfy the anti-monotone property, because of overlapping, and hence the number of the instances increases by the size of the pattern. Morimoto (2001) dealt with this case by using a constraint to obtain the same benefit as the anti-monotone property. The constraint was *"Any point object must belong to only one instance of a k-neighboring class set"*. Based on this constraint, he encountered problems obtaining different support for the same $k - neighboring$ class-set. The reason for the problem is that the selectivity process counted the instances from the same feature as one instance. If the order of the instances changed, then both the support value and the co-location patterns will change. Morimoto's method does not allow the generation of complex rules because of that constraint.

Huang et al. (2003) defined co-location patterns as the presence of a spatial feature in the neighborhood of instances of other spatial features. The authors developed an algorithm for mining valid rules in spatial databases, using an Apriori-based approach. Their algorithm does not separate the co-location mining and interesting pattern mining steps. The authors did not consider complex relationships or patterns, because they were pruning most items on the basis of their prevalence measure; this is known as the "Participation Index"; however, these items might contribute to forming complex rules in spatial mining. In contrast, our proposed technique, presented in Chapter 3, allows some items to be redundant as one of the reasons to form complex rules. Our approach distinguishes co-location mining from interesting pattern mining steps.

Munro et al. (2003) used cliques (a group of spatial objects located close to each other) as a co-location pattern. Similar to our approach, they separated clique mining from the pattern mining stages. However, they did not use maximal cliques (cliques not part of

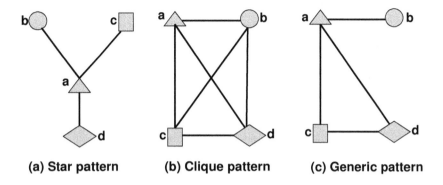

(a) Star pattern   **(b) Clique pattern**   **(c) Generic pattern**

Figure 2.4: A plot of different spatial co-location patterns.

another clique). They treated each clique as a transaction and used an Apriori-based technique for mining association rules. Since they used cliques (rather than maximal cliques)
as their transactions, the counting of pattern instances was very different. They considered complex relationships within the pattern mining stage. However, their definition
of negative patterns was very different – they used infrequent types, while we base our
definition on the concept of absence in maximal cliques. They used a different measure:
maxPI.

Arunasalam et al. (2005) used a similar approach to that of Munro et al. (2003). They
proposed an algorithm called NP_maxPI that used the MaxPI measure. Their proposed
algorithm prunes the candidate itemsets using a property of maxPI. They used an Apriori-
based technique to mine complex patterns. A primary goal of their work was to mine
patterns which have low support and high confidence. As with the work of Munro et al.
(2003), they did not use maximal cliques.

Zhang et al. (2004) enhanced the algorithm proposed in Huang et al. (2003) and used it
to mine special types of co-location relationships – *spatial stars* – in addition to cliques

and *generic* patterns (Figure 2.4). They used grid structure over space and a distance threshold ϵ to indicate the spatial relation between objects; that is, they located the object's neighbors by extending its coordinates by ϵ to form a disk. All grids that intersect with this disk will be hashed and all items inside these grids will be checked using the Euclidean distance measure to ensure they are close to the center of the star. Although their method for finding star patterns was based on two steps – a hashing and a mining step – they were insufficient to find clique-type patterns.

2.3.2 Spatial Challenges

Most of the work performed in the area of spatial databases has considered problems which treat spatial objects as lines, rectangles or points. These types are a challenge, given that special techniques are usually required to deal with them.

Since spatial data is fundamentally different from conventional data, no conventional technique can solve spatial problems. For example, the aggregation queries that retrieve the sum, count, and average from the conventional data are unusable in spatial data. These queries need to be modified, since the spatial objects consider the region features (Tao et al., 2004).

The shape of the spatial object plays a role as one of the challenges. In other words, taking the Minimum Bounding Rectangle (**MBR**) of the object leads approximately to the size of the object, which will be always bigger than the exact size. Analyzing spatial data is considered to be a big challenge, because objects' information may influence the objects that locate in the same region (Kriegel, 2005; Mamoulis et al., 2004).

2.4 Spatio-Temporal Data

This section gives an overview of spatio-temporal (ST) data. ST data contains the evolution of objects over time as well as their spatial features (Tsoukatos and Gunopulos, 2001). As with conventional databases, ST data can be used for retrieving information through queries. However, the retrieval process of information from data containing both spatial and temporal features is more complex. Tao et al. (2004) described the spatial query as one that focuses on retrieving data within boundaries, such as regions, districts and areas. For example, *Find the number of cars present in a district.* On the other hand, temporal queries focus on retrieving data within a time interval (i.e., a query interval). For example, the query *Find the traffic volume in a district during the past two hours.* One potential use of ST data is to store representations of interesting visual events as data records, such as moving body parts or human objects performing an activity (Kollios et al., 2001).

A Global Position System (GPS) is one example of a system that produces ST data. GPSs are used for mapping out objects' movements by storing their dynamic attributes, whose values change continuously over time (Wolfson et al., 1998). Geographic Information System (GIS) researchers compile the requirements for a ST information system, which merges the temporal and spatial features of the objects. This raises many issues, such as the components of change – the metrics, topologies and attributes of geographic objects – that may or may not change over time. Roshannejad and Kainz (1995) defined these changes as the "W-triangle" (where–what–when), which shows the component of the ST object. Abraham and Roddick (1999) reproduced eight combinations of these changes (Figure 2.5).

ST data stores the time interval (lifespan) of discrete phenomena. It contains the valid

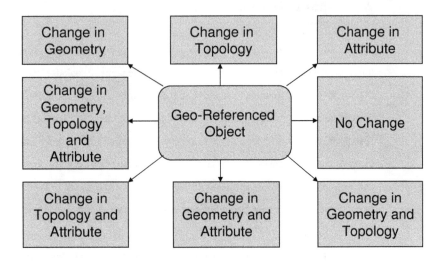

Figure 2.5: The eight possible spatio-temporal changes.

and transaction times explicitly. The data should support objects with identifiers to distinguish the investigated objects from each other; these identifiers play the role of the primary keys in the database.

Abraham and Roddick (1997) explained that extending spatial data to become ST data by adding the temporal element yields two kinds of rules: ST evolution rules, which demonstrate the changes of the objects over time; and ST Meta-rules, which describe the changes between the rules. The following section sheds light on the some of the interesting applications that produce ST data.

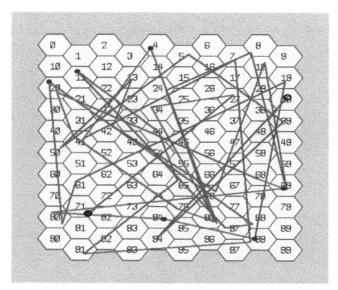

Figure 2.6: Two mobile-phone users' movements over 500 timesteps (Taheri and Zomaya, 2005).

2.4.1   Spatio-Temporal Applications

ST applications use information that describes moving objects within a district or region (*spatial feature*) during a given time interval (*temporal feature*). Many real-life
applications naturally provide huge amounts of data that contains a mixture of spatial
and temporal features; for instance, traffic supervision and analysis, human and animal
trajectories, and mobile services (Cheqing et al., 2004; Tao et al., 2004), land management, weather monitoring, and the tracing of mobile devices (e.g. tracking mobile-phone
users' trajectories) – Figure 2.6. Another example is the tracking the shapes of forests,
an application that stores information about the region of a forest as an object (spatial
coverage) that might change between time intervals (temporal coverage) due to natural
phenomena such as bushfires or drought (Laube and Imfeld, 2002; Lopez and Moon,

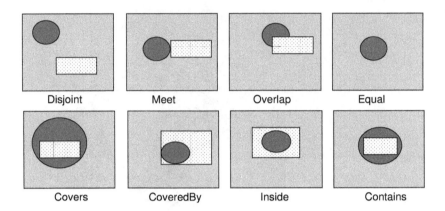

Figure 2.7: Relationships between two objects in 2-D space.

2005). Traffic management, sensor networks, and stock control are further examples of ST applications (Verhein and Chawla, 2008).

As mentioned previously, ST data is a combination of objects' spatial attributes (points, lines, rectangles) and temporal attributes (events, interval). Abraham and Roddick (1999) demonstrated that besides the temporal and spatial features and conventional attributes, there are relationships between ST objects that provide useful information. These relationships describe the real-life interaction between objects (Figure 2.7).

Lopez and Moon (2005) defined ST objects as objects with spatial and temporal extents. The spatial extent includes points, rectangles and lines; the temporal extent can be either the valid time or the transaction time for objects.

ST objects are described as moving objects (i.e., those that change their location over time). Cao et al. (2005) defined the objects' movement (i.e., trajectories) as ordered sequences of spatial sites sampled at multiple time stamps.

The most common type of ST data consists of movement traces of point objects. The widespread availability of GPS-enabled mobile devices and location-aware sensors has led to an increase in the generation and availability of ST data. A unique example of a project that is continuously generating ST data is related to the tracking of caribou in Northern Canada. Since 1993, the movement of caribou has been tracked through the use of GPS collars with the underlying expectation that the data collected will help scientists to understand the migration patterns of caribou and help them to locate their breeding and calving locations (pch, 2007). While the number of caribou tagged at a given time is small, the time interval the temporal data for each animal is long. One of the major challenges in ST query processing and patten discovery is to efficiently handle "long duration" ST data (Chapter 4). Interesting ST queries can be formulated on this particular dataset. For example, *How are herds formed and does herd membership change over time? Are there specific regions in Northern Canada where caribou herds tend to rendezvous?*

2.4.2 Motion Patterns

Moving objects have special attributes that describe their behavior (motion). For example, each object should comprise ID, coordinates to indicate its location, and a timestamp for each location – a combination of normal (ID), spatial (coordinates), and temporal (timestamp) features describe the object's movement. Most of the applications that store information about moving objects produce huge amounts of daily data, especially if the number of the monitored objects is substantial.

Laube and Imfeld (2002) described the motion pattern as a defined set of motion parameters that could be valued either over time or across objects. Hence, three types of patterns exist: patterns of the same object defined over several time intervals $(t, 1)$ are called *sequences*; patterns across more than one object at the same timestamp $(1, n)$, which are called *incidents*; combined patterns over time and across objects, i.e., a combination of several time intervals and several objects (t, n), which are called *interactions*. Motion patterns are described as "Group patterns". Where, for example, these patterns are a group of users, intuitively the motion of these users is a series of information points containing spatial and temporal dimensions (Wang et al., 2003).

2.4.3 Spatio-Temporal Mining

The focus in data mining research is to design techniques to discover new patterns in large repositories of ST data. For example, Mamoulis et al. (2004) mine periodic patterns moving between objects and Ishikawa et al. (2004) mine ST patterns in the form of Markov transition probabilities. More recently Verhein and Chawla (2008) have proposed efficient association mining-type algorithms to discover ST patterns such as sinks, sources, stationary region, and thoroughfares. The emphasis in what this book will propose is not to mine new unknown patterns, but to efficiently query existing complex ST-patterns.

The problem of detecting movement patterns in ST data has recently received considerable attention from several research communities, e.g., geographic information science (Gudmundsson et al., 2008; Shirabe, 2006), data mining (du Mouza and Rigaux, 2005; Jeung et al., 2008; Kollios et al., 2001; Koubarakis et al., 2003; Verhein and Chawla, 2008), databases (Asakura and Hato, 2004; Güting et al., 2003, 2000; Güting and Schneider, 2005; Park et al., 2003; Wolfson and Mena, 2004; Wolfson et al., 1998), and algorithms (Buchin et al., 2008a,b; Gudmundsson et al., 2007). One of the first movement patterns studied (Jensen et al., 2007; Jeung et al., 2007, 2008; Kalnis et al., 2005; Laube and Imfeld, 2002) was moving clusters. Unsurprisingly, this was one of the first patterns to be studied since it is the ST equivalent of point clusters in a spatial setting. A moving cluster in a time interval T consists of at least m entities, such that for every point within T there is a cluster of m entities. The set of entities might be much larger

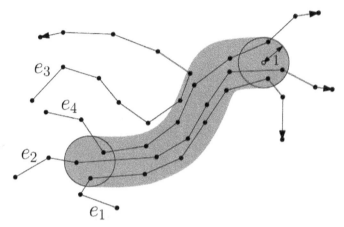

Figure 2.8: A flock pattern among four trajectories. If $m = 3$ and the radius $r = 1$ then the longest-duration flock lasts for six time steps. Entities e_1, e_2 and e_4 form a flock, since they move together within a disk of radius r, while e_3 is excluded in this pattern. Here, the flock uses Definition 2.1.

than m, thus entities may join and leave a cluster during the cluster's lifetime. A moving cluster is sometimes called a *variable subset flock*. Closely related to moving clusters is the *flock* pattern, or *fixed subset flock* (Gudmundsson and van Kreveld, 2006). This problem has been studied in several papers (Benkert et al., 2006; Gudmundsson and van Kreveld, 2006; Gudmundsson et al., 2007; Jeung et al., 2007; Laube and Imfeld, 2002; Laube et al., 2004). Even though different papers use slightly different definitions, the main idea is that a flock consists of a fixed set of entities moving together as a cluster during the duration of the flock pattern.

More recently, further movement patterns have been studied. Jeung et al. (2008) modified the definition of a flock to what they call a "convoy", formed from a group of entities that are density-connected. Intuitively, two entities in a group are density-connected if a sequence of objects exists that connects the two objects, and the distance between consecutive objects does not exceed a given constant.

In this book we will focus on the fixed subset flock pattern. Laube and Imfeld (2002) proposed the REMO framework (RElative MOtion), which defines similar behavior within groups of entities. They define a collection of ST patterns based on similar direction of motion or change of direction. Laube et al. (2004) extended the framework by not only including direction of motion, but the location itself. They defined several ST patterns, including flock (moving close together in coordination), leadership (spatially leading the movement of other objects), convergence (converging towards a spot), and encounter (gathering at a spot), and designed algorithms to compute them efficiently.

However, these algorithms only consider each time step separately; that is, given $m \in \mathbb{N}$ and $r > 0$, a flock is defined by at least m entities within a circular region of radius r and moving in the same direction at a given point in time. Benkert et al. (2006) argued that this is insufficient for many practical applications – e.g., a group of animals may need to stay together for days or even weeks before it is defined as a flock. They proposed the following definition of a flock that takes the minimum duration (k) into account (Figure 2.8); m, k and r are given as parameters, and are hence fixed values.

Definition 2.1 (m, k, r)-*flock$_A$ - Given a set of n trajectories where each trajectory consists of τ line segments, a flock in a time interval $I = [t_i, t_j]$, where $j - i + 1 \geq k$, consists of at least m entities, such that for every point in time within I there is a disk of radius r that contains all of the m entities, and $m, k \in \mathbb{N}$ and $r > 0$ are given constants.*

Benkert et al. (2006) proved that there is an alternative, algorithmically simpler definition of a flock that is equivalent provided that the entity moves with constant velocity along the straight line between two consecutive points.

Benkert et al. (2006) described efficient approximation algorithms for reporting and detecting flocks, where the size of the region is permitted to deviate slightly from what is specified. Approximating the size of the circular region within a factor of $\Delta > 1$

means that a disk with radius between r and Δr that contains at least m objects may or may not be reported as a flock, while a region with a radius of at most r that contains at least m entities will always be reported. Their main approach is a $(2 + \varepsilon)$-approximation with running time $T(n) = O(\tau n k^2 (\log n + 1/\varepsilon^{2k-1}))$. Even though the dependency on the number of entities is small, the dependency on the duration of the flock pattern is exponential. Thus, the only significant open problem that remains in Benkert et al. (2006) is to develop a robust algorithm with a smaller dependency on k. This is exactly our focus in this book, and will be discussed thoroughly in Chapter 4.

2.4.4 Spatio-Temporal Challenges

Although many real-life applications generate ST data, the data is unattainable for research purposes, primarily because of the need for data privacy; for this reason, researchers tend to design their own (synthetic) data.

Even though ST data can be obtained from some applications, it needs to be transformed and designed to suit the mining process. This transformation typically requires more time than generating synthetic data does. This has been another common reason that researchers have designed synthetic data to represent their desired format.

ST data size is a vital challenge, it makes the mining process infeasible. An example of an application that generates massive data is mobile device applications. These record the location of mobile users periodically. For example, if the location of a user is stored every 60 seconds, the number of stored locations over a year is 525600 locations. Generated data involving this many locations is described as a high-dimensional, and can be significantly larger depending on the time interval and how frequent the location is stored.

Using the spatial and temporal features for an object at the same time is complex because of the striking differences in their features. One of these main differences is that time has natural ordering feature, but space does not (Verhein and Chawla, 2008).

2.5 The Curse of Dimensionality

We now consider the situation when data has been collected and is in the form shown in Array 2.1. In matrix notation, the data is described as $X_{n \times d}$, where n is the number of described objects and d is the number of features (dimensions).

$$X = \begin{pmatrix} x_{11} & x_{12} & \cdots & x_{1d} \\ x_{21} & x_{22} & \ddots & x_{2d} \\ \vdots & \vdots & \ddots & \vdots \\ x_{n1} & x_{n2} & \cdots & x_{nd} \end{pmatrix} \tag{2.1}$$

For example, one of the interesting features of microarray experiments is the fact that they gather information on a large number of genes (for example, 6000). If we consider genes as variables, this means that our observations are in a 6000-dimensional space and $d >> n$ (where d is the number of dimensions and n is the number of gene expressions). Another example of high-dimensional data is in Section 2.4, that is ST data. For example, one of the applications that produce ST data is tracking cell-phone users' trajectories. This application generates an enormous number of data points that describe the users' locations. Therefore, if we consider each location as a variable (dimension), this yields a massive number of dimensions (Example 2.1).

Example 2.1 *If we intend to track 100 mobile users for three months, and the location is stored every second for each user. That means our raw data ($X_{n \times d}$) will have 100 rows, and the number of dimensions will consist of three months of data, that is –*

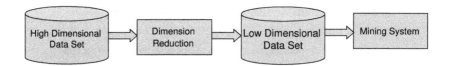

Figure 2.9: Dimensionality reduction process. The mining process is only effective with vector data of not more than a certain number of dimensions. Hence, high-dimensional data must be transformed into low-dimensional data before it is used in the mining system.

$$
\begin{aligned}
d &= 3 \times 30 \; days \\
  &= 3 \times 30 \times 24 \; hours \\
  &= 3 \times 30 \times 24 \times 60 \; minutes \\
  &= 3 \times 30 \times 24 \times 60 \times 60 \; seconds \\
  &= 7776000.
\end{aligned}
$$

As a result, our matrix will be of size 100×7776000, which is a massive number of dimensions.

Example 2.1 shows the "curse of dimensionality", a term coined by (Bellman, 1961). Such high dimensionality brings the need for a dimensionality reduction process to (i) reduce the dimension of the data to a manageable size, (ii) keep as much of the original information as possible, and (iii) feed the reduced dimension data into the system. Figure 2.9 summarizes this situation, showing the dimension reduction as a preprocessing stage in a data mining system.

2.5.1   Dimensionality Reduction Techniques

Dimensionality reduction is a process of mapping the original data into lower-dimensional space while retaining the structure of the original data as much as possible. It can be applied to solve the curse of dimensionality, introducing only minimal distortion.

Several techniques are available for carrying out the dimensionality reduction, e.g., PCA (Principal Components Analysis), Discrete Fourier Transform (DFT), Discrete Wavelet Transform (DWT), Single Value Decomposition (SVD), and random projection (RP) (Achlioptas, 2003). This section presents two techniques (PCA, RP) and briefly describes several of their uses. We will not study the other techniques (SVD, DFT, and DWT) given that they are all ranking techniques, similar to PCA.

2.5.1.1   Principle Components Analysis (PCA)

PCA (Principle Components Analysis) is a well-known dimensionality reduction technique. It searches for the vectors that best describe the data in a minimum squared error sense. PCA can be calculated in several ways, including finding the eigenvectors of the sample covariance matrix across all classes.

The way PCA operates is as follows. It rotates the coordinate system, "drops" some of the dimensions (axes), and effectively projects the data into a lower-dimension space. The new coordinate system is determined based on the direction of the variance in the data. For most real datasets, most of the variance is captured in the first few new dimensions; by "dropping" the other dimensions, while preserving the "essential structure" of the data. PCA does not guarantee the preservation of the pairwise distance (even approximately) between data points. Bishop (2006) is a good source for a geometric and algebraic introduction to PCA.

PCA can be used to project an $m \times n$ matrix A into a rank $\kappa < min(m, n)$ matrix A_κ, which minimizes the Frobenius norm[1] $||A - C||_F$ over all matrices of rank κ. Since our objective is to project points into a lower-dimensional space to efficiently use spatial data structures and preserve pairwise distances, PCA may not be particularly useful. The time complexity of PCA is $O(mn^3)$.

Algebraically, PCA works as follows. Let A be an $m \times n$ matrix, where m is the number of objects, each of dimension n. Then A can be decomposed as

$$A = \Sigma_{i=1}^r \sigma_i u_i v_i^t.$$

Here, $\sigma_i's$ are the eigenvalues of the mean-centered covariance matrix $A^t A$, $v_i's$ are the corresponding eigenvectors, r is the rank of the matrix A, and $u_i's$ are the new coordinates in space spanned by the new eigenvectors. The decomposition of A is organized in a way such that $\sigma_1 > \sigma_2 > \ldots > \sigma_r$. Geometrically, the $\sigma_i's$ are variance in the direction determined by the $v_i's$. Thus, for $k < r$,

$$A' = \Sigma_{i=1}^k \sigma_i u_i v_i^t$$

is effectively a lower-dimension projection of the objects in A.

PCA has been used by Hotelling (1933) as the best-known statistical analysis technique for detecting network traffic anomalies. Recent papers in the networking literature have applied it to the problem of traffic anomaly detection with promising initial results (Lakhina et al., 2004b,a; Ringberg et al., 2007).

PCA has been used as a popular technique for parameterizing shape, appearance and motion (Cootes et al., 2001; Black et al., 1997; Moghaddam and Pentland, 1995; Murase

---

[1]It is also known as the Euclidean norm.

and Nayar, 1995). These learned PCA representations have proved to be valuable for solving problems such as face and object recognition and background modeling (Cootes et al., 2001; Moghaddam and Pentland, 1995; Oliver et al., 2000; Belhumeur et al., 1997; Witten and Frank, 2000) .

Malagón-Borja and Fuentes (2005) presented an object detection system to detect pedestrians in gray-level images, without assuming any prior knowledge of the image. The system works as follows: in the first stage, a classifier based on PCA examines and classifies each location in the image at different scales; in the second stage, the system eliminates all false detections.

2.5.1.2   Random Projections (RP)

Random projection (RP) has recently emerged as a powerful method for dimensionality reduction. In Example 2.1, our data matrix $n \times d$, where d, is huge. In random projection, the original d-dimensional data is projected to a κ-dimensional ($\kappa << d$) subspace through the origin, using a random $\kappa \times d$ matrix R whose columns have a given unit lengths.

Intuitively, the data will be projected into a new space X' using a basic matrix multiplication operation; this is given in Equation 2.2.

$$Projected\ Data\ =\ Original\ Data \times Random\ Matrix$$
$$X'\ =\ X_{n \times d} \times R_{d \times \kappa} \tag{2.2}$$

The key idea of random mapping arises from the Lemma by (Johnson and Lindenstrauss,

1982). If points in a vector space are projected onto a randomly selected subspace of suitably high dimension, then the distances between the points are approximately preserved.

The choice of the random matrix R is one of the key points of interest. The elements r_{ij} of R are often Gaussian distributed $N(0,1)$. Achlioptas (2003) has recently refined the Johnson-Lindenstrauss Lemma. He proposed sparse random projections by replacing the $N(0,1)$ entries in R with entries in $\{-1, 0, 1\}$ with probabilities $\{\frac{1}{6}, \frac{2}{3}, \frac{1}{6}\}$ accomplishing a threefold increase in processing time. In other words, Achlioptas (2003) refined the Johnson-Lindenstrauss Lemma. The random matrix entries will be defined as:

$$r_{ij} = \sqrt{3} \times \begin{cases} +1 & \text{with probability} \quad 1/6 \\ 0 & \quad .. \quad\quad\quad 2/3 \\ -1 & \quad .. \quad\quad\quad 1/6. \end{cases}$$

The sparse random projections method (Achlioptas, 2003) has recently become popular. It was first applied and experimentally tested on image and text data by (Bingham and Mannila, 2001). Random projections have been used in Machine Learning (Arriaga and Vempala, 1999; Bingham and Mannila, 2001; Fern and Brodley., 2003; Fradkin and Madigan, 2003; Kaski, 1998) – more specifically, in the analysis of phenomena such as Latent Semantic Indexing (Vempala, 1998; Tang et al., 2004), finding motifs in biosequences (Buhler and Tompa, 2001), face recognition (Goel et al., 2005), and privacy preserving distributed data mining (Liu and Ryan, 2006). Random projections have been used in a number of contexts in recent high-dimensional geometric constructions and algorithms (Johnson and Lindenstrauss, 1982; Frankl and Maehara, 1987; Chor and Sudan, 1998; Kleinberg, 1997). For example, Kleinberg (1997) built a data structure from the projection of a set P onto random lines through the original data.

Random projections are used in the area of data clustering. Kaski (1998) showed that the inner product (the similarity) between the mapped (projected) vectors closely follows

Dim Reduction Technique	Length Limitation	Computational cost	Space cost	Indexable
PCA	No	$O(dN^3)$	$O(dN)$	Yes
RP	No	$O(kdN)$	$O(kN)$	Yes

Table 2.2: Comparison of dimensionality reduction techniques (PCA and RP).

the inner product of the original vectors. He showed that the random mapping method is computationally feasible choice for dimensionality reduction in circumstances where the reduced dimensional data vectors are used for clustering or similar approaches; this is especially the case if the original dimensionality of the data is very large and it is infeasible to use more computationally expensive methods such as PCA.

Urruty et al. (2007) proposed a new method of clustering using random projections. Their algorithm consists of two phases: a projection phase (which creates uni-dimensional histograms and aggregates these histograms to produce the initial clusters) and the second phase that consists of certain post-processing techniques of clusters obtained by several random projections. Their experiments show the potential of the algorithm as a segmentation technique and provide a reliable criterion for the validation of clusterings[2].

Li et al. (2006) have proposed new probabilities for random matrix entries, in which they recommended using R of entries in $\{-1, 0, 1\}$ with probabilities $\{\frac{1}{2\sqrt{d}}, 1 - \frac{1}{\sqrt{d}}, \frac{1}{2\sqrt{d}}\}$ for accomplishing a significant $\sqrt{d}$-fold speedup, with only a small loss in accuracy.

Incorporating RP in our method – which reports long durations and a complex ST pattern named "flock" – has solved the dimensionality burden with high levels of accuracy (Chapter 4).

---

[2]Clusterings: is another synonym of the clustering process.

2.5.2 Comparison of Dimensionality Reduction Techniques

Table 2.2 provides a comparison between two dimensionality reduction techniques (PCA and RP) when applying them on a dataset ($X_{n \times d}$), where n is the number of described objects and d is the number of dimensions. The comparison shed light on important features of these techniques, such as whether they have a limitation on length, computation cost, and space cost, and whether they are indexable.

2.6 Time Series Data Mining

This section provides an introduction to time series data and the similarity measures used when mining time series data.

Definition 2.2 *A time series is a sequence of data points, which are measured typically at successive times, where each data point represents a value. For example, a time series (sequence) S is described as follows:*

$$S = \{s_t : t \in \mathcal{N}\}, \tag{2.3}$$

where s_t represents a data point at time t.

Definition 2.3 *A time series data TSD is a collection of sequences (trajectories), which is described as follows:*

$$TSD = \{S_i : i \in \mathcal{N}\}, \tag{2.4}$$

where S_i is the ith sequence.

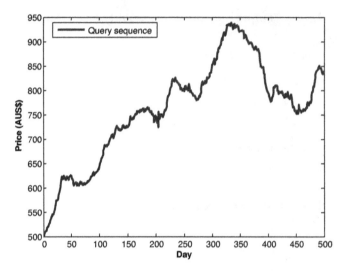

Figure 2.10: A query sequence from a stock time series dataset.

An example of a time series (sequence) is shown in Figure 2.10; the sequence represents the movement of stock price over 500 days. To distinguish between time series data and trajectory (ST) data, there are two differences:

1. Most time series are one-dimensional data where $d = 1$, whereas, trajectory data is often two or more dimensions $d \geq 2$.

2. In many applications the time dimension is very important. In ST data, the time feature of a trajectory is important for answering special queries that involve "time slice" or "time interval" (Pfoser et al., 2000; Tao and Papadias, 2001; Xu et al., 1990).

As mentioned in Chapter 1, one of the focuses in this book is on time series shape similarity. Thus, throughout the book, the time component of the time series and trajectories data is disregarded when measuring the similarity. More examples about application that produce time series data will be provided in Chapters 5 and 6.

2.6.1 Time Series Similarity Measures

This section sheds light on these measures, focusing on previous work which used three famous distance measures: Euclidean distance, Dynamic Time Warping, and Longest Common Subsequence.

Definition 2.4 *Given a data space $\mathcal{D}$ defined on time series (trajectory) data and any two data points $x, y \in \mathcal{D}$, a distance function, **dist**, on $\mathcal{D}$ is defined as:*

$$\textbf{dist} : \mathcal{D} \times \mathcal{D} \longrightarrow \mathbb{R} \tag{2.5}$$

*where $\mathbb{R}$ denotes the set of real numbers. The distance function **dist** is known as a metric if it satisfies the following properties:*

- *Nonnegativity: distance is positive between two different points.*

$$\textbf{dist}(x, y) \geq 0. \tag{2.6}$$

- *Reflexivity: distance is zero precisely from a point to itself.*

$$\textbf{dist}(x, y) = 0 \text{ if and only if } x = y. \tag{2.7}$$

- *Symmetry: distance between x and y is the same in either direction.*

$$\textbf{dist}(x, y) = \textbf{dist}(y, x). \tag{2.8}$$

- *Triangle inequality: distance between two points is the shortest distance along any path.*

$$dist(x, y) \leq dist(x, z) + dist(z, y). \tag{2.9}$$

Definition 2.5 *Given a data space $\mathcal{D}$, $x, y \in \mathcal{D}$, distance function **dist** on $\mathcal{D}$, and a predefined threshold ϵ: x and y are considered similar if **dist**$(x, y) \leq \epsilon$, and dissimilar if **dist**$(x, y) > \epsilon$.*

The similarity between time series is achieved by matching the points in the first time series with the corresponding points in the second. A distance function and a predefined threshold, as mentioned in Definition 2.5, should be used to measure the similarity. The distance function is application- and data-dependent, and needs to be carefully designed to meet application requirements.

2.6.2 Euclidean Distance (EucDist)

Euclidean distance is the first distance function that was used as similarity measure in time series database literature (Agrawal et al., 1993; Faloutsos et al., 1994; Rafiei and Mendelzon, 1998). The definition of the Euclidean distance depends on the type of the data – dimensionality of the data indicates which equation we should use when calculating the distance. Therefore, there are two main definitions:

Definition 2.6 *For any two 1-dimension points, $S = (s_x)$, and $Q = (q_x)$, the Euclidean distance is defined as follows:*

$$
\begin{aligned}
EucDist &= \sqrt{(s_x - q_x)^2} \\
&= |s_x - q_x|.
\end{aligned}
\tag{2.10}
$$

Definition 2.7 *EucDist: Given two n-dimensions sequences* $Q = (q_1, \ldots, q_n)$ *and* $S = (s_1, \ldots, s_n)$ *where* $n > 1$ *and denotes to the length of the two sequences, the Euclidean distance between them is:*

$$
\begin{aligned}
EucDist &= \sqrt{(q_1 - s_1)^2 + \cdots + (q_n - s_n)^2} \\
&= \sqrt{\sum_{i=1}^{n} (q_i - s_i)^2}.
\end{aligned}
\tag{2.11}
$$

2.6.2.1 Euclidean Distance Limitations

Two of the main advantages of the EucDist are that it is easy to compute and the computational cost is linear in terms of the time series length. However, EucDist has limitations that make it unsuitable for similarity mining in time series data: it requires sequences to be of the same length, and does not support local time shifting. Local time shifting occurs when one element of one sequence is shifted along the time axis to match an element of another time sequence (even when the two matched elements appear in different positions in the sequences). This is useful when the sequences have similar shape but are out of phase. It is called "local", because not all of the elements of the query sequence need to be shifted and the shifted elements do not have the same shifting factor. By contrast, in "global" time shifting, all of the elements are shifted along the time axis by a fixed shifting factor. More specifically, local time shifting cannot be handled by EucDist, because EucDist requires the ith element of query sequence to be aligned with the ith element of the data sequence. Figure 2.11(a) shows an example on how elements from two time series could be matched (aligned) when EucDist is computed in the presence of local time shifting. These limitations of the Euclidean distance are the reasons for using another similarity measure – the Dynamic Time Warping (DTW) measure, which will be

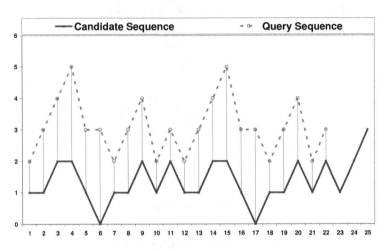

(a) Using Euclidean distance. EucDist=11.2458

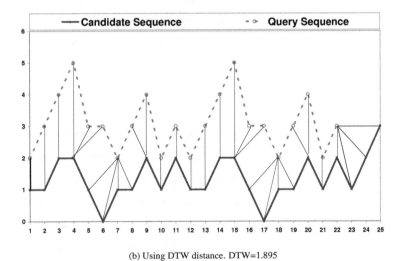

(b) Using DTW distance. DTW=1.895

Figure 2.11: Point correspondence when two time series contains local time shifting.

introduced in the following section.

2.6.3 Dynamic Time Warping (DTW)

Dynamic time warping (DTW) is a dynamic programming technique for measuring the similarity between any two sequences (i.e., two time series) which may differ in time or speed. For example, similarities in walking patterns would be detected, even if in one video the person was walking slowly and in another was walking quickly, or even if there were accelerations and decelerations during the course of the video.

2.6.3.1 DTW Features

DTW has valuable features: it does not require that the two sequences being compared are of the same length, and can handle local time shifting (Figure 2.11(b)). However, is not considered as a metric, because it does not satisfy the triangular inequality. Theorem 2.1 state this and we give a proof for our theorem by a counter example.

Recall the definition of the triangular inequality in Definition 2.4. Here, we restate it when dealing with time series data as in Definition 2.8.

Definition 2.8 *Given three time series Q, S and R. The distance function **dist**, defined on a data space $\mathcal{D}$, satisfies the triangular inequality if and only if Inequality 2.12 holds.*

$$\forall\, Q, R, S \in \mathcal{D},\ \textbf{\textit{dist}}(Q, R) + \textbf{\textit{dist}}(R, S) \geq \textbf{\textit{dist}}(Q, S) \tag{2.12}$$

Theorem 2.1 *DTW is not a metric because it does not satisfy the triangular inequality which is given in the Inequality 2.12.*

Proof Given three time series data, $Q = \{1\}$, $R = \{1, 0, 2\}$ and $S = \{2, 13, 4\}$.

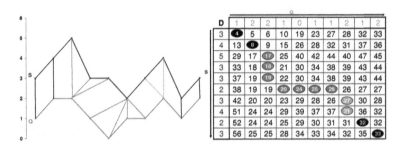

(a) Warping S and Q where the lines be-   (b) The warping matrix D produced by
tween the points are the warping costs.   DTW; highlighted cells constitute the op-
                                                 timal warping path.

Figure 2.12: Aligning two time series S and Q using *DTW*. The lines between the points are
the warping costs.

$DTW(Q, R) = 2$, $DTW(R, S) = 7$, and $DTW(Q, S) = 14$. It is clear that $DTW(Q, S) >$
$DTW(Q, R) + DTW(R, S)$. This violates the triangular inequality.   □

Chapter 5 provides more explanation about the DTW constraints and provides its pseudocode.

2.6.3.2 DTW Calculation

In general, DTW is a technique designed to find an optimal match between any two given
sequences with certain constraints. The sequences are "warped" non-linearly in the time
dimension (Figure 2.12). Definition 2.9 provides more about how DTW works.

Definition 2.9 DTW: *Is a dynamic programming technique which divides the problem
into several sub-problems, each of which contribute in calculating the distance between
any two sequences cumulatively. Equation 2.13 shows the recursion that governs the
computation:*

$$D(i,j) = d(i,j) + min \begin{cases} D(i-1,j) \\ D(i-1,j-1) \\ D(i,j-1). \end{cases} \tag{2.13}$$

Example 2.2 *Given two sequences (S, Q) each of which consists of 10 data points:*

$$S = \{3, 4, 5, 3, 3, 2, 3, 4, 2, 3\}$$

and

$$Q = \{1, 2, 2, 1, 0, 1, 1, 2, 1, 2\}.$$

Figure 2.12(a) illustrates the process of warping S and Q, where the lines between the points are the warping costs. By applying the Equation 2.13, the warping matrix D is produced (Figure 2.12(b)); highlighted cells constitute the optimal warping path between S and Q.

The previous example showed the warping process between two sequences of the same length. Definition 2.10 gives a general formula for the DTW distance between any two sequences.

Definition 2.10 *The general definition of DTW distance between two time series S and Q of lengths M and N, respectively, is defined as:*

$$DTW(S,Q) = \begin{cases} \infty & if\, M = 0 \; or \; N = 0 \\ 0 & if\, M = 0 \; and \; N = 0 \\ \textbf{dist}(s_1, q_1) + min \begin{pmatrix} DTW(S', Q') \\ DTW(S', Q) \\ DTW(S, Q') \end{pmatrix} & otherwise. \end{cases} \qquad (2.14)$$

where $\textbf{dist}(s_1, q_1)$ is the EucDist as defined in Equation 2.10. S' and Q' are the remaining subsequences of S and Q, respectively.

2.6.3.3   DTW Applications

DTW has been applied to many applications, such as video (Blackburn and Ribeiro, 2007) and graphics (Chen et al., 2004). Any data which has a linear representation can be analyzed using DTW. A well-known application has been automatic speech recognition (Itakura, 1975; Tappert and Das, 1978; Myers et al., 1980; Sakoe and Chiba, 1978; Sankoff and Kruskal, 1999; Rabiner and Juang, 1993), enabling it to cope with different speaking speeds.

In bioinformatics, Aach and Church (2001) successfully applied DTW to RNA expression data. DTW has been effectively used to align biometric data, such as signatures (Munich and Perona, 1999) and even fingerprints (Kovacs-Vajna, 2000).

DTW was first introduced to the data mining community in the context of mining time series (Berndt and Clifford, 1994). Since it is a flexible measure for time series similarity, it is used extensively for ECGs (Electrocardiograms) (Caiani et al., 1998), speech processing (Rabiner and Juang, 1993), and robotics (Schmill et al., 1999).

2.6.3.4 DTW Complexity

The time and space complexities of DTW in the best case are both quadratic, $O(MN)$, where M and N are the lengths of the two time series being compared. Therefore, when the dataset size increases, a large amount of time is required when computing DTW to answer a particular query. This can be solved in two ways: by using indexing methods to reduce the number of computations required, and by speeding up the technique used to calculate the DTW itself. Several techniques have been introduced to speed up *DTW* and/or reduce the space overhead (Hirschberg, 1975; Berndt and Clifford, 1996; Yi et al., 1998; Kim et al., 2001; Keogh and Ratanamahatana, 2004).

2.6.3.5 Speeding up DTW

This section explains the state-of-the-art techniques used to speed up DTW. The techniques are grouped into three different categories – adding constraints on DTW, lower bounding, and approximating techniques.

1. **Adding Constraints on DTW.**

   Many researchers have added constraints on DTW to increase its speed by limiting how far the warping path may stray from the diagonal of the warping matrix (Tappert and Das, 1978; Berndt and Clifford, 1994; Myers et al., 1980). The division of matrix that the warping path is allowed to visit is called the warping window. Two popular global constraints are Itakura Parallelogram and Sakoe-Chiba Band (Itakura, 1975; Sakoe and Chiba, 1978); they used to speed up the *DTW* by adding constraints which force the warping path to lie within a band around the diagonal; if the optimal path crosses the band, the DTW distance will not be optimal. Figures 2.13(a) and 2.13(b) illustrate the Itakura Parallelogram and Sakoe-Chiba bands, respectively.

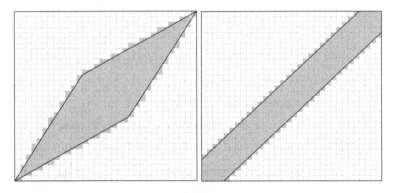

(a) Global constraint (Itakura Parallelo-   (b) Global constraint (Sakoe Chiba Band).
gram).

Figure 2.13: Illustration of the two well known global constraints used on DTW. The global constraint limits the warping scope. The diagonal gray areas correspond to the warping scopes.

2. **Lower Bounding DTW.**

Kim et al. (2001) introduced a lower bound function which extracts only four features from each sequence. The features are the first and the last of the elements of the sequence, together with the maximum and minimum values in the same sequence. The lower bound of DTW distance between candidate and query sequences is calculated as the maximum squared difference of corresponding features extracted from both sequences. Figure 2.14 illustrates this technique.

Yi et al. (1998) introduced another lower bound function which exploits the observation that all points in the query sequence are larger than the maximum of the candidate sequence (or smaller than the minimum). The squared difference of these values is the final DTW distance. Figure 2.15 illustrates this technique.

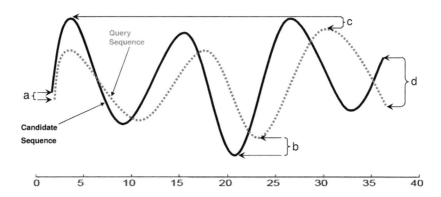

Figure 2.14: The lower bound introduced by (Kim et al., 2001). The squared difference is calculated from the difference in first points (a), last points (d), minimum points (b) and maximum points (c).

Keogh and Ratanamahatana (2004) introduced an efficient lower bounding technique, LB_Keogh. It reduces the number of DTW computations in a time series database context; however, it does not reduce the space complexity of the DTW computation. The idea is to use a cheap lower bounding calculation, which will allow us to do the expensive calculations when are they essential. Equation 2.15 shows the LB_Keogh between a query Q and a candidate C sequence. Figure 2.16 depicts two sequences (Q,C) as well as the *lower* (L) and *upper* (U) bound of Q. After finding the lower and upper bounds, the distance is calculated based on Equation 2.15. LB_Keogh can be treated similarly as the **MBR** index, since the two bounds (Upper and Lower) appear as an envelope around the query sequence; this is similar to the idea of having a rectangle (box) around the spatial object.

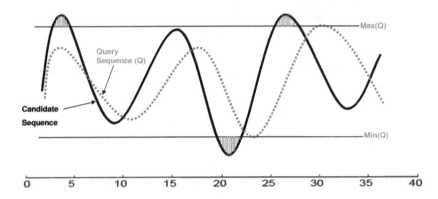

Figure 2.15: An illustration of the lower bound introduced by (Yi et al., 1998). The sum of the squared length of the gray lines is the over all DTW distance.

$$LB\_Keogh(Q,C) = \sqrt{\sum_{i=1}^{n} \begin{cases} (c_i - U_i)^2 & \text{if } c_i > U_i \\ (c_i - L_i)^2 & \text{if } c_i < L_i \\ 0 & otherwise. \end{cases}} \qquad (2.15)$$

The LB_keogh function can be readily visualized as the Euclidean distance between any part of the candidate matching sequence not falling within the envelope and the nearest (orthogonal) corresponding section of the envelope (Figure 2.16).

DTW has been used in data streaming problems. Capitani and Ciaccia (2007) proposed a new technique, Stream-DTW (*STDW*). This measure is a lower bound of the *DTW*. Their method uses a sliding window of size 512. They incorporated a band constraint, forcing the path to stay within the band frontiers, as in (Sakoe and Chiba, 1978).

Sakurai et al. (2005) presented FTW, a search method for DTW; it adds no global constraints on DTW. Their method designed based on a lower bounding distance

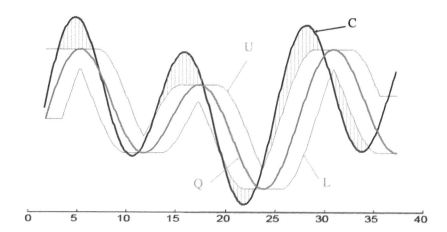

Figure 2.16: An illustration of the lower bound LB_Keogh, where C is the candidate sequence, Q is the query sequence, U and L are the upper and lower bounds of Q, respectively. The sum of the squared length of the gray lines is the overall DTW distance. The figure is adopted from (Keogh and Ratanamahatana, 2004). This lower bound used Sakoe-Chiba Band.

measure that approximates the DTW distance. Therefore, it minimizes the number of DTW computations but does not increase the speed the DTW itself.

3. **Approximating DTW.**

Salvador and Chan (2007) introduced an approximation algorithm for *DTW* called *FastDTW*. Their algorithm begins by using *DTW* in very low resolution, and progresses to a higher resolution linearly in space and time. *FastDTW* is performed in three steps: coarsening shrinks the time series into a smaller time series; the time series is projected by finding the minimum distance (warping path) in the lower resolution; and the warping path is an initial step for higher resolutions. The authors refined the warping path using local adjustment. *FastDTW* is an approximation algorithm, and thus there is no guarantee it will always find the optimal path. It requires the coarsening step to be run several times to produce many different resolutions of the time series. The *FastDTW* approach depends on a radius parameter as a constraint on the optimal path; however, our technique does not place any constraints while calculating the *DTW* distance.

2.6.4   Longest Common Subsequence (LCSS)

The LCSS problem is defined as finding the longest common subsequence to all sequences in a dataset of sequences.

LCSS is a matching measure proposed to handle a special case of time series data, when the data may contain possible noise (Hirschberg, 1975; Boreczky and Rowe, 1996). The noise could be introduced by hardware failures, disturbance signals, or transmission errors. The intuition of LCSS is to remove the noise effects by counting only the number of matched elements between two time series. The LCSS definition is given in Definition 2.11.

Definition 2.11 *The general definition of LCSS measure between two time series S and Q of lengths M and N, respectively, is defined as:*

$$
LCSS(S,Q) \;=\; \begin{cases} \infty & if\, M = 0 \, or \, N = 0 \\ 0 & if\, M = 0 \, and \, N = 0 \\ LCSS(S',Q') + 1 & if\; \mathbf{dist}(s_1, q_1) \leq \epsilon \\ max \begin{pmatrix} LCSS(S',Q) \\ LCSS(S,Q') \end{pmatrix} & otherwise. \end{cases} \tag{2.16}
$$

where $\mathbf{dist}(s_1, q_1)$ is the EucDist as defined in Equation 2.10. S' and Q' are the remaining *subsequences of S and Q, respectively. ϵ is the matching threshold.*

We will not use this measure, because our datasets are noise-free data. Our problem is to find the similarity between time series, not the longest common subsequence. The reason we introduced LCSS was merely to show the difference between it and the other measures.

Distance measure	Different lengths	Local time shifting	Noise	Setting matching threshold	Computational cost	Metric
EucDist	No	No	No	No	$O(N)$	Yes
LCSS	Yes	Yes	Yes	Yes	$O(N^2)$	No
DTW	Yes	Yes	No	No	$O(N^2)$	No

Table 2.3: Comparison between three different distance measures on raw representation of time series data.

2.6.5   Comparison between Similarity Measures

Table 2.3 gives a comparison between the three distance measures, based on six criteria: the ability to handle sequences with different lengths; the ability to handle sequences with local time shifting; the ability to handle sequences that contain noise; whether a matching threshold is needed; computational cost; and whether the distance function is a metric. From Table 2.3, the following remarks can be made:

- The computational cost of DTW is quadratic, and it is not metric distance measure; it is not possible to improve the retrieval efficiency with distance-based access methods. The same applies to LCSS.

- DTW does handle time sequences with different lengths and local time shifting, but EucDist does not.

- LCSS is not sensitive to noisy data. DTW and EucDist are both sensitive but the datasets that used in this research are noise-free data.

- LCSS needs a predefined matching threshold to handle noisy data.

2.7 Summary and Conclusions

This chapter has reviewed the state-of-the-art research in three areas: spatial, spatio-temporal and time series data. We summarize them as follows:

- In the area of spatial data , we have provided an overview about spatial applications and the challenges encountered when mining spatial data.

- This chapter provides most of the interesting work in the area of ST data that has been conducted recently. More specifically, it describes the work that was performed on the moving objects problems. We described the term curse of dimensionality as a challenge that made many of the proposed techniques do not work efficiently. Two dimensionality reduction approaches and the differences between them were explored.

- An introduction about time series data has been given. A thorough explanation about three different similarity measures was provided. The three measures include: Euclidean, DTW, and LCSS distances. This chapter provides a comparison between these similarity measures as well as the most interesting applications that have used them recently.

Chapter 3

Mining Complex Co-location Rules
(MCCRs)

Following a comprehensive summary of the related work, this chapter proposes a heuristic (**GridClique**) to mine efficiently the maximal clique patterns from a large spatial dataset. It combines the proposed algorithm with an association rule mining technique to efficiently generate complex co-location rules. The chapter is organized as follows: Section 3.1 provides an introduction, and Section 3.2 places our contributions in the context of related work. Further details of our proposed method (MCCR) are provided in Section 3.3. Our experiments and the analysis of the results are explained in Section 3.4, and Section 3.5 concludes the chapter. Table 3.1 lists all notations used in this chapter[1].

---

[1]This chapter is based on the following publications:

- Ghazi Al-Naymat. **Enumeration of Maximal Clique for Mining Spatial Co-location Patterns**. Proceedings of the 6th ACS/IEEE International Conference on Computer Systems and Applications (AICCSA), Doha, Qatar. Mar 31st – Apr 4th, 2008. Pages (126–133) (Al-Naymat, 2008).

- Florian Verhein and Ghazi Al-Naymat. **Fast Mining of Complex Spatial Co-location Patterns using GLIMIT**. The 2007 International Workshop on Spatial and Spatio-temporal Data Mining (SSTDM) in cooperation with The 2007 IEEE International Conference on Data Mining (ICDM). Omaha NE, USA. October 28–31, 2007. Pages (679–684) (Verhein and Al-Naymat, 2007).

Symbol	Description
MCCR	Mining Complex Co-location Rule
parsec	Unit of length used in astronomy. It stands for "**par**allax of one arc **sec**ond"
Mpc	An abbreviation of "Mega-parsec", which is one million parsecs, or 3261564 light years
arcmin	Unit of angular measurement. Sizes of objects on the sky, field of view of telescopes, or practically any angular distance "Arc of Minutes"
minPI	Minimum Participation Index
maxPI	Maximum Participation Ratio
SDSS	Sloan Digital Sky Survey
O	Spatial objects with fixed locations
o_i	The ith spatial object
G	Undirected graph
C_M	Maximal Clique
$g(\cdot)$	Function that performs a transformation on the transposed dataset
$\circ$	An operator that combined two *itemvectors* together to create a new *itemvector* corresponding to the union of the two itemsets
P	A set of complex types
$N(\cdot)$	Number of maximal cliques that contain the set of complex types
T	Set of transactions
I	Item set
$f(\cdot)$	Item set size
$|\cdot|$	Number of elements in a set
$Card(\cdot)$	Cardinality of a set
w/o	Without
w	With
X	X coordinate
Y	Y coordinate
$\lceil \cdot \rceil$	Function that gives the ceiling of a value
$\lfloor \cdot \rfloor$	Function that gives the floor of a value
EucDist	Euclidean distance

Table 3.1: Description of the notations used.

3.1 Introduction

Spatial data is essentially different from transactional data in its nature. The objects in a spatial database are distinguished by a spatial (location) and several non-spatial (aspatial) attributes. For example, an astronomy database that contains galaxy data may contain the

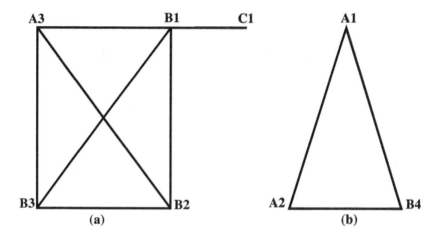

Figure 3.1: An example of clique patterns.

x, y and z coordinates (spatial features) of each galaxy, their types and other attributes.

Spatial datasets often describe *geo-spatial* or *"astro-spatial"* (astronomy related) data. In this work, we use a large astronomical dataset containing the location of different *types* of galaxies. Datasets of this nature provide opportunities and challenges for the use of data mining techniques to generate interesting patterns. One such pattern is the *co-location* pattern. A co-location pattern is a group of objects (such as galaxies) each of which is located in the neighborhood (within a given distance) of another object in the group.

A *clique* is a special type of co-location pattern. It is described as a group of objects such that *all* objects in that group are co-located with each other. In other words, given a predefined distance, if a group of objects lie within this distance from every other object in the group, they form a clique. Figure 3.1 shows eight different objects $\{A1, A2, A3, B1, B2, B3, B4, C1\}$. The set $\{B1, B2, A3\}$ is a clique. However, $\{B1, B2, A3, C1\}$ is not, because $C1$ is not co-located with $B2$ and $A3$, therefore $\{B1, B2, A3, C1\}$ is a co-location pattern only.

ID	Maximal Cliques	Raw Maximal Cliques	Non-Complex Relationships	Complex Without Negative Relationships	Complex With Negative Relationships
1	{A3, B1, B2, B3}	{A, B, B, B}	{A, B}	{A, B, B+}	{A, B, B+, -C}
2	{B1, C1}	{B, C}	{B, C}	{B, C}	{-A, B, C}
3	{A1, A2, B4}	{A, A, B}	{A, B}	{A, A+, B}	{A, A+, B, -C}

Table 3.2: Representing maximal cliques of Figure 3.1 as complex relationships

In this chapter we consider *maximal cliques*. A maximal clique is a clique that does not appear as a subset of another clique in the same co-location pattern (and therefore the entire dataset, as each object is unique). For example, in Figure 3.1, $\{A1, A2, B4\}$ forms a maximal clique as it is not a subset of any other clique. However, $\{A3, B2, B3\}$ is not a maximal clique since it is a subset of the clique $\{A3, B1, B2, B3\}$ (which in turn *is* a maximal clique). The second column of Table 3.2 shows all the maximal cliques in Figure 3.1.

In our dataset, each row corresponds to an object (galaxy) and contains its type as well as its location. We are interested in mining relationships between the *types* of objects. Examples of object types in this dataset are "early-type" galaxies and "late-type" galaxies. To clarify, we are not interested in the co-locations of *specific* objects, but rather, the co-locations of their *types*. Finding complex relationships between such types is useful information in the astronomy domain – this will be shown in Section 3.1.2. In Figure 3.1, there are three types: $\{A, B, C\}$.

In this chapter we focus on using maximal cliques to allow us to mine interesting *complex spatial relationships* between the object types. A *complex spatial relationship* includes not only whether an object type, say A, is present in a (maximal) clique, but:

- Whether *more than one* object of its type is present in the (maximal) clique. This is a *positive type* and is denoted by $A+$.

- Whether objects of a particular type are not present in a *maximal clique* – that is, the absence of types. This is a *negative type* and is denoted by $-A$.

The inclusion of *positive* and / or *negative types* makes a relationship *complex*. This allows us to mine patterns that state, for example, that A occurs with multiple Bs but not with a C. That is, the presence of A may imply the presence of multiple Bs and the absence of C. The last two columns of Table 3.2 show examples of (maximal) complex relationships. We propose an efficient algorithm (GridClique) to extract all maximal clique patterns from a large spatial data – the Sloan Digital Sky Survey (SDSS) data.

We are not interested in *maximal* complex patterns (relationships) in themselves, as they provide only local information (that is, about a maximal clique). We are however interested in *sets* of object types (including complex types) that appear across the entire dataset (that is, among many maximal cliques). In other words, we are interested in mining *interesting complex spatial relationships* (sets), where "interesting" is defined by a global measure. We use a variation of the $minPI$ (minimum Participation Index) measure (Shekhar and Huang, 2001) to define interestingness (Equation 3.1).

$$minPI(P) = \min_{t \in P}\{N(P)/N(\{t\})\} \qquad (3.1)$$

Here P is a set of complex types we are evaluating, and $N(\cdot)$ is the number of maximal cliques that contain the set of complex types. We count the occurrences of the pattern (set of complex types) only in the maximal cliques. This means that if the $minPI$ of a pattern is above α, then we can say that whenever any type $t \in P$ occurs in a maximal clique, the entire pattern P will occur at least as a fraction α of those maximal cliques. Using $minPI$ is superior to simply using $N(P)$ because it scales by the occurrences of the individual object types, thus reducing the impact of a non-uniform distribution on the object types.

In this work we focus on *maximal cliques* because:

- The process of forming complex positive relationships makes sense. Suppose we extract a clique that is not maximal, such as $\{A1, B4\}$ from Figure 3.1. We would not generate the positive relationship $\{A+, B\}$ from this, even though each of $\{A1, B4\}$ are co-located with $\{A2\}$. So we get the correct pattern only once we have considered the maximal cliques.

- Negative relationships are possible. For example, consider the maximal clique in row 1 of Table 3.2. If we did not use *maximal* cliques, then we would consider $\{B1, B2, B3\}$, and from this we would *incorrectly* infer that the complex relationship $\{B, B+, -A\}$ exists. However, this is not true because A *is* co-located with each of $\{B1, B2, B3\}$. *Therefore, using non-maximal cliques will generate* incorrect *negative patterns.*

- Each maximal clique will be considered as a single instance (transaction) for the purposes of counting. In other words, we automatically avoid multiply counting the same objects within a maximal clique.

- Mining maximal cliques reduces the number of cliques by removing all redundancy. It is possible to mine for maximal cliques directly. And because negative types cannot be inferred until the maximal clique is mined, it does not make sense to mine cliques that are not maximal.

The previous reasons demonstrate the value of our proposed algorithm GridClique given it mines all maximal clique patterns from large spatial data and puts them into a format which can be mined easily using association rule mining techniques.

3.1.1   Problem Statement

The problem that we are considering in this chapter consists of two parts:

1. **Given a large spatial dataset (astronomy dataset), extract all maximal clique patterns. More specifically, extract all co-location patterns that are not subsets of any other co-location patterns.**

2. **Given the set of maximal cliques, find all interesting and complex patterns that occur among the set of maximal cliques. More specifically, find all *sets* of object types, including *positive* and *negative* (that is, complex) types that are interesting as defined by their $minPI$ being above a threshold.**

To solve the above problem efficiently, we propose a heuristic based on a divide-and-conquer strategy. Our proposed heuristic is GridClique (Al-Naymat, 2008) as it will be described in Section 3.3.1. After obtaining the maximal clique patterns the problem therefore becomes an *itemset mining* task. To achieve this very quickly, we use the GLIMIT algorithm (Verhein and Chawla, 2006) as we will describe in Section 3.3.3.

Including negative types makes the problem much more difficult, as it is typical for spatial data to be sparse. This means that the absence of a type can be very common. Approaches relying on an Apriori style algorithm find this very difficult, however, this is not a problem for our approach.

3.1.2   Contributions

In this chapter we make the following contributions:

1. We introduce the concept of *maximal cliques*. We describe how the use of *maximal*

Does the presence of

Elliptical Galaxy
(Early)

Spiral Galaxy
(Late)

Figure 3.2: An interesting question which can be answered by our method.

cliques makes more sense than simply using cliques, and we show that they allow the use of negative patterns.

2. We propose a heuristic GridClique on the basis of a divide-and-conquer strategy, to mine maximal clique patterns that will be used in mining complex co-location rules.

3. We show that GLIMIT can be used to mine complex, interesting co-location patterns very efficiently in very large datasets. We demonstrate that GLIMIT can be almost three orders of magnitude faster than using an Apriori based approach.

4. We propose a general procedure that splits the maximal clique generation, complex pattern extraction and interesting pattern mining tasks into modular components.

5. We contribute to the astronomy domain by proving existing facts when analyzing the complex association rules that are generated by our proposed approach. For

example, we managed to answer questions such as the one depicted in Figure 3.2.

3.2   Related Work

Huang et al. (2003) defined the co-location pattern as the presence of a spatial feature in the neighborhood of instances of other spatial features. They developed an algorithm for mining valid rules in spatial databases using an Apriori-based approach. However, their algorithm does not separate the co-location mining and interesting pattern mining steps as our approach does, and they did not consider complex relationships or patterns.

Munro et al. (2003) used cliques as a co-location pattern. Similarly to our approach, they separated the clique mining from the pattern mining stages. However, they did not use maximal cliques. They treated each clique as a transaction and used an Apriori-based technique for mining association rules. Since they used cliques (rather than maximal cliques) as their transactions, the counting of pattern instances is different. They considered complex relationships within the pattern mining stage. However, their definition of negative patterns is very different – they used infrequent types while we base our definition on the concept of absence in *maximal cliques*. They used a different measure, namely, maximum Participation Ratio (maxPI).

Arunasalam et al. (2005) used a similar approach to Munro et al. (2003). They proposed an algorithm (NP_maxPI) which used the MaxPI measure. The proposed algorithm prunes the candidate itemsets using a property of maxPI. They used an Apriori-based technique to mine complex patterns. A primary goal of their work was to mine patterns which have low support and high confidence. As with the work of Munro et al. (2003), they did not use maximal cliques.

Zhang et al. (2004) enhanced the algorithm proposed in Huang et al. (2003) and used

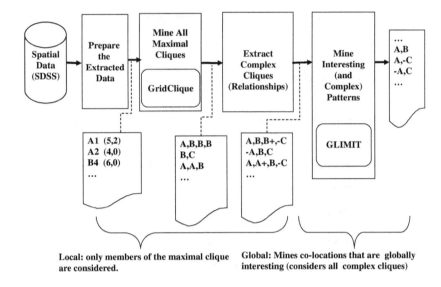

Figure 3.3: The complete process of Mining Complex Co-location Rules (MCCRs).

it to mine special types of co-location relationships in addition to cliques, namely the *spatial star*, and *generic* patterns.

To the best of our knowledge, previous work has used Apriori-type algorithms for mining interesting co-location patterns. We use GLIMIT (Verhein and Chawla, 2006) as the underlying pattern mining algorithm; this will be discussed in Section 3.3.3. To the best of our knowledge, no previous work has used the concept of *maximal cliques*.

3.3   Mining Complex Co-location Rules (MCCRs)

The MCCR process is illustrated through the flowchart in Figure 3.3. It consists of four stages. First, the data preparation process, which starts by downloading the spatial data from the SDSS repository and then putting it in the desired format. Appendix A provides comprehensive details about this stage. Second, the GridClique algorithm finds all *maximal clique patterns*, and strips them of the object identifiers – producing raw maximal cliques (Table 3.2). The GridClique algorithm is described in Section 3.3.1. One pass is then made over the raw maximal cliques in order to extract complex relationships. We describe this in Section 3.3.2. This produces maximal complex cliques, each of which is then considered as a *transaction*. An *interesting itemset mining algorithm*, using *minPI* as the interestingness measure, is used to extract the interesting complex relationships. We describe this in Section 3.3.3.

Figure 3.3 shows that the clique generation and complex relationship extraction are local procedures, in the sense that they deal only with individual maximal cliques. In contrast, the interesting pattern mining is global – it finds patterns that occur across the entire space. We consider subsets of maximal cliques only in the last step – after the complex patterns have been extracted.

3.3.1   Mining Maximal Cliques

The stage of mining maximal cliques is the process of transforming the raw spatial data into transactional type data that can be used by any association rule mining techniques. This is a crucial step and it is performed using the our proposed algorithm (GridClique).

3.3.1.1   Basic Definitions and Concepts

Consider a set of objects O with fixed locations. Given an appropriate distance measure $d : O \times O \rightarrow \mathbb{R}$ we can define a graph G as follows; let O be the vertices and construct an edge between two objects $o_1 \in O$ and $o_2 \in O$ if $d(o_1, o_2) \leq \tau$, where τ is a chosen distance. A *co-location pattern* is a connected subgraph.

Object type	X-Coordinate	Y-Coordinate
A1	2.5	4.5
A2	6	4
A3	2	9
B1	1.5	3.5
B2	5	3
B3	5	4
C1	2.5	3
C2	6	3
D1	3	9
D2	7	1.5

Table 3.3: An example of two-dimensional dataset.

Definition 3.1 (Clique) *A clique $C \in O$ is any fully connected subgraph of G. That is,* $d(o_1, o_2) \leq \tau \ \forall \{o_1, o_2\} \in C \times C.$

As we have mentioned in Section 3.1, we use maximal cliques so that we can define and use complex patterns meaningfully and to avoid double counting.

Definition 3.2 (Maximal Clique) *A maximal clique C_M is a clique that is not a subset (sub-graph) of any other clique. So that $C_M \not\subset C \ \forall C \in O$.*

Definition 3.3 (Clique Cardinality) *A cardinality ($Card$) is the size of a clique and it is given in Equation 3.2. For example, if we have a clique $C = \{o_1, o_2, o_3\}$, then $Card(C) = 3$.*

$$Card(C) = |\{o \in O : o \in C\}|, \tag{3.2}$$

where $| \cdot |$ denotes the number of elements in a set.

Generally, cardinality of a set is a measure of the "number of elements of the set".

3.3.1.2 GridClique Algorithm

The GridClique algorithm uses a heuristic based on a divide-and-conquer strategy to efficiently extract maximal clique patterns from large spatial dataset (SDSS). That is achieved by dividing the space into a grid structure based on a predefined distance. The use of the grid structure plays a vital role for reducing the search space. Our heuristic treats the spatial objects (galaxies) as points in a plane and it uses grid structure when mining the maximal clique patterns. We use the Euclidean distance as the distance measure, because it is very efficient to compute.

The aim of the GridClique algorithm is to extract maximal clique patterns that exist in any undirected graph. It is developed using an index structure through grid implementation. Table 3.3 contains 10 objects and their X and Y coordinates; this information will be used to explain the functionality of the GridClique algorithm. The SDSS is a three-dimensional dataset, but in our example we use two-dimensional dataset for simplicity. Algorithm 3.1 displays the pseudocode of the GridClique algorithm.

The ten spatial objects in Table 3.3 are depicted in Figure 3.4. The figure will be used as an example when we explain our algorithm. The objects are used as spatial points and are placed in the plane using their coordinates. The edges in each subgraph are generated between every two objects on the basis of the co-location condition. That is, if the distance between any two objects is $\leq d$, where d is a predefined threshold, an edge will be drawn between the two objects (Figure 3.4(a)). The GridClique algorithm works as follows.

1. It divides the space into a grid structure which contains cells of size $d \times d$ (Lines 1 to 12). The grid space is structured where each cell has a key (GridKey). This key is a composite of X, Y and Z coordinates (Line 5). We used a special data structure (hashmap) which is a list that stores data based on an index (key) to speed

---

Algorithm 3.1 GridClique algorithm.

Input: Set of points $(P_1, \cdots, P_n)$, Threshold d
Output: A list of maximal clique patterns.
    {**Generating grid structure.**}
1: $GridMap \leftarrow \phi$
2: $PointList \leftarrow \{P_1, \cdots, P_n\}$
3: **for all** $P_i \in PointList$ **do**
4:     Get the coordinates of each point Pk_x, Pk_y, Pk_z
5:     Generate the composite key (GridKey=(Pk_x, Pk_y, Pk_z)).
6:     **if** $GridKey \in GridMap$ **then**
7:         $GridMap \leftarrow P_i$
8:     **else**
9:         $GridMap \leftarrow$ new GridKey
10:       $GridMap.GridKey \leftarrow P_i$
11:    **end if**
12: **end for**
    {**Obtaining the neighborhood lists.**}
13: **for all** $p_i \in GridMap$ **do**
14:     $p_i.list \leftarrow \phi$
15:     $NeighborGrids \leftarrow$ (the 27 neighbor cells of p_i)
16:     $NeighborList \leftarrow \phi$
17:     **if** $NeighborGrids_i.size() > 1$ **then**
18:         **for all** $p_j \in NeighborGrids_j$ **do**
19:            **if** EucDist $(p_i, p_j) \leq d$ **then**
20:              $p_i.list \leftarrow p_j$ (p_i, p_j are neighbors)
21:            **end if**
22:         **end for**
23:     **end if**
24:     $NeighborList \leftarrow p_i.list$
25: **end for**
    {**Pruning neighborhood list if at least one of its items violates the maximal clique definition.**}
26: $TempList \leftarrow \phi$
27: $MCliqueList \leftarrow \phi$
28: **for all** $Record_i \in NeighborList$ **do**
29:     $RecordItems \leftarrow Record_i$
30:     **for all** $p_i \in RecordItems$ **do**
31:         **for all** $p_j \in RecordItems$ **do**
32:            **if** $EucDist(p_i, p_j) \leq d$ **then**
33:              $Templist \leftarrow p_j$ (p_i, p_j are neighbors)
34:            **end if**
35:         **end for**
36:     **end for**
37:     $MCliqueList \leftarrow Templist$
38: **end for**

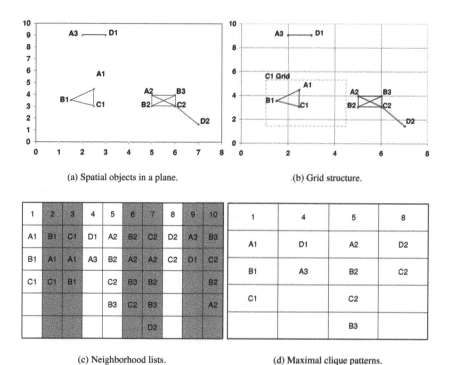

(a) Spatial objects in a plane.

(b) Grid structure.

(c) Neighborhood lists.

(d) Maximal clique patterns.

Figure 3.4: An example to illustrate the process of extracting maximal clique patterns from 2 dimensions dataset.

up retrieving the stored data. The number of cells depends on the maximum value of X and Y coordinates in the dataset. Having coordinates for each cell, helps in placing the points in their corresponding cells. The total number of cells is calculated using Equation 3.3.

$$Number\ of\ Cells = \lceil \frac{max(X)}{d} \rceil \times \lceil \frac{max(Y)}{d} \rceil, \tag{3.3}$$

where $max(X)$ and $max(Y)$ are the maximum of X and Y coordinates, respectively. The function $\lceil \cdot \rceil$ is the ceiling of any value.

In our example data, the maximum values of X and Y coordinates are 7 and 9, respectively. The predefined distance $d = 2$. Using Equation 3.3, the number of cells will be 20. After structuring the grid space, the algorithm then places the points into their corresponding cells (Lines 6 to 11). This is performed by considering the X and Y coordinates of the corresponding cell as the $\lfloor X \rfloor^2$ and $\lfloor Y \rfloor$ coordinates of the placed point (Figure 3.4(b)). For example, if we consider object $\{A1\}$, its X and Y coordinates are 2.5 and 4.5, respectively; to place it in the grid space, its corresponding cell will be the one which has $GridKey = (2, 4)$.

2. GridClique finds (Lines 13-25) each object's neighbors and adds them to a list – this list is the neighborhood list. The algorithm checks the neighborhood list members with respect to the definition of maximal clique which is all members (objects) are co-located with each other. In other words, the distance between every two objects is $\leq d$. For each object, the algorithm considers 9 cells[3] (NeighborGrids in Line 15) to check their members if they are co-located with the object being checked.

Considering the example in Figure 3.4, a list for each object is created. Our concern is to find only co-location patterns that have at least two objects (i.e., cardinality ≥ 2), because one object does not give co-location information. Therefore, there is no need to count objects that do not have connections (i.e., a relationship) with at least one other object. However, in our example all objects share relationships with others. For example, object $\{A1\}$ has a relationship with objects $\{B1,C1\}$. It can be seen that these objects share the same neighborhood (co-located) – a neighborhood list will be generated for object $\{A1\}$. Figure 3.4(c) shows the neighborhood

---

[2]$\lfloor \cdot \rfloor$ is a function that gives the floor of a value.
[3]In two-dimensional space the number of neighbor cells is 9, however, in three-dimensional space it is 27.

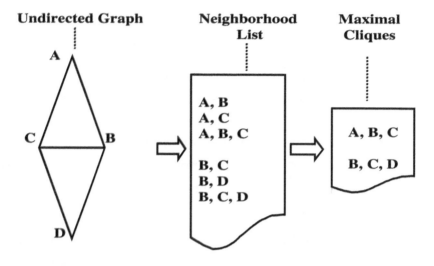

Figure 3.5: Undirected graph contains two maximal cliques.

list for each object.

3. It prunes any neighborhood list that contains at least one object that violates the maximal clique definition (Lines 26-38). For example, list 7 is pruned because one of its members $\{A2\}$ is not co-located with $\{D2\}$. The shaded lists (2, 3, 6, 9, and 10) in Figure 3.4(c) are pruned for the same reason. The $MCliqueList$ (Line 27) is a special data structure called *hashset*. This set does not allow repetition – this has helped us to report only distinct maximal clique patterns.

As a result of the previous steps, a list of distinct maximal clique patterns will be reported. For example, $\{A1, B1, C1\}$ is a maximal clique, and so forth for lists 4, 5 and 8 (Figure 3.4(d)).

3.3.1.3 GridClique Algorithm Analysis

This section discusses the GridClique algorithm completeness, correctness, and complexity.

Completeness: All points in neighborhood lists appear as set or subset in maximal clique lists. After acquiring the entire neighbors for each point, another check among these neighbors is conducted to assure that all points are neighbors with each other. Intuitively, doing that results to have repeated neighborhood lists. Therefore, this ensures finding all maximal cliques in any given graph. The repeated neighborhood lists will be pruned using the *hashset* data structure.

Correctness: Every subset of a maximal clique appears in the neighborhood list. Thus, all maximal cliques that appear in maximal clique lists will not be found as a subset in another maximal clique, since this is the definition of maximal clique. Figure 3.5 displays an undirected graph and the neighborhood list and the existing maximal clique patterns. As can be seen, the pair $\{A, D\}$ does not appear in the neighborhood list, because the distance between A and D does not satisfy the co-location condition. Therefore, the pair $\{A, D\}$ will not be included in the maximal cliques list. In other words, any subset of any maximal clique that appears in the neighborhood list will not appear as an independent maximal clique. In this way, the correctness of the proposed algorithm is shown.

Complexity: Firstly, assume there are N points and c cells, and assume that all points are uniformly distributed. Hence, on average there is N/c points per cell. Assume each cell has l neighbors. Then to create the neighborhood list of one point, $l(N/c)$ points need to be examined to check if they are within distance d. Since the total number of points is N, the cost is $O(N^2 l/c)$. And since $c >> l$, an assumption, that this part of the algorithm is sub-quadratic, can be stated.

Secondly, the pruning stage for the neighborhood lists. Again assume that on average the

Relationship	Notation	Description	Example
Non-Complex	$A \rightarrow B$	Presence of B in the neighborhood of A.	Sa type spiral galaxies $\rightarrow$ Sb type spiral galaxies.
Positive	$A \rightarrow A+$	Presence of many instances of the same feature in a given neighborhood	Elliptic galaxies tend to cluster more strongly. E$\rightarrow$ E+.
Negative	$A \rightarrow -B$	Absence of B in the neighborhood of A.	Elliptic galaxies tend to exclude spiral galaxies. E$\rightarrow$-S.
Complex	$A+ \rightarrow -C, B$	Combination of two or more of the above relationships.	Clusters of elliptic galaxies tend to exclude other types of galaxies. E+$\rightarrow$-S.

Table 3.4: Spatial relationships with real-life examples from the astronomy domain.

length of each neighborhood list is k. Then for each neighborhood list, k points have to be checked against the co-location condition – the cost is $O(k^2)$. The total cost for this step is $O(Nk^2)$.

Ultimately, the total cost is the cost of putting the points in cell (O (N)), the cost of creating the neighborhood lists $O(N^2l/c)$, and the cost of pruning the lists $O(Nk^2)$. Therefore, the complexity of the algorithm is $O(N(Nl/c + k^2 + 1))$.

3.3.2 Extracting Complex Relationships

A relationship is complex if it consists of *complex types* as defined in Section 3.1.

Extracting a complex relationship R from a maximal clique C_M is straightforward – we simply use the following rules for every type t:

1. If C_M contains an object with type t, $R = R \cup t$ (non-complex relationship).

2. If C_M contains more than one object of type t, $R = R \cup t+$ (positive relationship).

3. If C_M does not contain an object of type t, $R = R \cup -t$ (negative relationship).

If R includes a positive type $A+$, it will *always* include the basic type A. This is necessary so that maximal cliques that contain $A+$ will be counted as containing A when we mine for interesting patterns.

As mentioned earlier, the negative type makes sense only if we use *maximal cliques*. The last three columns of Table 3.2 show the result of applying Rule 1, Rule 1 and Rule 2, and all three rules, respectively. Table 3.4 provides four relationship types supported with real-life examples from the astronomy domain.

3.3.3   Mining Interesting Complex Relationships

In *itemset mining*, the dataset consists of a set of transactions T, where each transaction $t \in T$ is a subset of a set of *items* I; that is, $t \subseteq I$. In our work, the set of complex maximal cliques (relationships) becomes the set of transactions T. The items are the object types – including the complex types such as $A+$ and $-A$. For example, if the object types are $\{A, B, C\}$, and each of these types is present and absent in at least one maximal clique, then $I = \{A, A+, -A, B, B+, -B\}$. An interesting itemset mining algorithm mines T for interesting itemsets. The support of an itemset $I' \subseteq I$ is the number of transactions containing the itemset: $support(I') = |\{t \in T : I' \subseteq t\}|$. So called *frequent itemset mining* uses the support as the measure of interestingness. For reasons described in Section 3.1 we use *minPI* (see Equation 3.1) which, under the mapping described above, is equivalent to Equation 3.4.

$$minPI(I') = \min_{i \in I'}\{support(I')/support(\{i\})\} \qquad (3.4)$$

Since $minPI$ is *anti-monotonic*, we can easily prune the search space for interesting patterns.

GLIMIT (Verhein and Chawla, 2006) is a very fast and efficient itemset mining algorithm that has been shown to outperform Apriori (Agrawal and Srikant, 1994) and FP-Growth (Han et al., 2000). GLIMIT works by first transposing the dataset, so that each row, known as an *itemvector*, corresponds to an item. GLIMIT then makes one pass over the result (the *itemvectors*). It is an *item enumeration* algorithm, which means that it searches through the space of possible itemsets. It does this in a bottom up (the size of the itemsets increases along a branch of the search) fashion, so is suitable for measures that possess a form of *anti-monotonic* property (Verhein and Chawla, 2006). The search progresses in a depth first fashion, which enables very little space to be used – specifically, space linear in the size of the dataset (Verhein and Chawla, 2006). GLIMIT uses a framework defined by the following functions and operator (Verhein and Chawla, 2006):

- $g(\cdot)$ performs a transformation on the transposed dataset.

- $\circ$ is an operator that combined two *itemvectors* together to create a new *itemvector* corresponding to the union of the two itemsets.

- $m_{I'} = f(\cdot)$ is a measure on an itemset $I' \subseteq I$ (evaluated over the corresponding *itemvector*) that depends only on that itemset.

- $M_{I'} = F(\cdot)$ is a measure on an itemset $I' \subseteq I$ that uses $f(\cdot)$ and may depend on said itemset as well as any of its subsets.

The *minPI* measure can be incorporated into GLIMIT as follows (let $I' = \{1, 2, ..., q\}$ for simplicity): $g(\cdot)$ is the identity function (there is no transformation on the dataset), $\circ = \cap$ and $f(\cdot) = |\cdot|$ (the set size). This means that $m_{I'} = support(I')$. Finally, $M_{I'} = F(m_{I'}, m_1, ..., m_q) = \min_{i \in I'}\{m_{I'}/m_i\}$.

We use GLIMIT with the above instantiations of its framework to mine interesting co-locations (Figure 3.3). For comparison, we will use an Apriori (Agrawal and Srikant, 1994) implementation.

The Apriori (Agrawal and Srikant, 1994) and Apriori-like algorithms are *bottom up item enumeration* type itemset mining algorithms. Apriori works in a breadth first fashion, making one pass over the dataset for each level expanded. This is in contrast to GLIMIT, which makes only one pass over the entire dataset. In Apriori, a *candidate generation* step generates candidate itemsets (itemsets that may be interesting) for the next level, followed by a dataset pass (*support counting*) where each candidate itemset is either confirmed as interesting, or discarded. The support counting step is computationally intensive as subsets of the transactions need to be generated. GLIMIT operates on a completely different principle (Verhein and Chawla, 2006).

3.4 Experiments and Results

3.4.1 Experimental Setup

All experiments were carried out on "Windows XP Pro" operated laptop with a 2.0GHz Pentium 4M processor and 2 GB main memory. The data structures and algorithms were implemented in Java.

We used a real life three-dimensional astronomy dataset from the Sloan Sky Digital Survey (SDSS)[4]. We extracted all galaxies from this dataset, giving a total of $365,425$ objects. There were 12 *types* of galaxies. The distance threshold used for generating the maximal cliques was 1 Mpc. Comprehensive details about the data preparation process are given in Appendix A.

A total of $121,506$ maximal cliques (transactions) were generated in 39.6 seconds. This is quite a large dataset. We processed these in a number of ways as described in Section 3.3.2:

---

[4]http://cas.sdss.org/dr6/en/tools/search/sql.asp

Maximal Clique Set	Items	Average Size (Transaction Width)
Non-Complex	12	1.87
Complex w/o Negative	21	2.69
Complex w Negative	33	13.69

Table 3.5: Description of the resulting sets of maximal cliques.

- **Non-Complex:** We removed duplicate items (object types) in the maximal cliques. For example, a set of object types $\{A, A, B\}$ becomes $\{A, B\}$.

- **Complex w/o Negative:** We included *positive* types: if an object type A occurred more than once in a maximal clique, we replaced it with A and $A+$. For example, a set of object types $\{A, A, B\}$ becomes $\{A, A+, B\}$.

- **Complex w Negative:** The same as **Complex w/o Negative**, but we included *negative* types. That is, we added all object types that were not present in the maximal clique as negative items. For example, a set of object types $\{A, A, B\}$ becomes $\{A, A+, B, -C\}$.

Table 3.5 describes the resulting sets of maximal cliques we used for mining interesting patterns.

The "Complex w Negative" dataset is very large. It has $121,506$ transactions (like the others), but each transaction has an average size of 13.7.

3.4.2 Results

This section reports the results obtained from the use of the GridClique algorithm and the process of mining co-location rules.

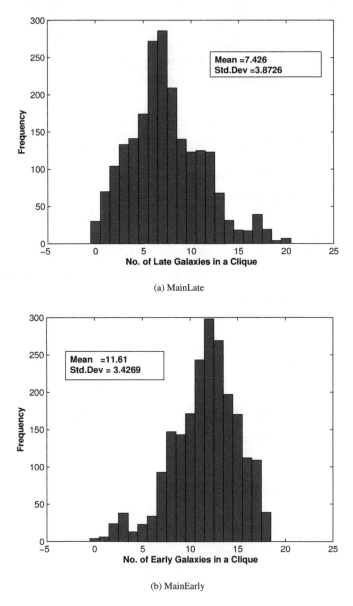

(a) MainLate

(b) MainEarly

Figure 3.6: The existence of galaxies in the universe.

3.4.2.1 Galaxy Types in Large Maximal Cliques

In this experiment we applied the GridClique algorithm on the "Main" galaxies[5] extracted from SDSS data to generate maximal cliques with neighborhood distance as 4 Mpc. We selected the cliques with the largest cardinality ($Card = 22$). Figures 3.6(a) and 3.6(a) show the distribution of "Late" and "Early" type galaxies in the reported cliques, respectively. These results show that large cliques consist of more "Early" type galaxies (Elliptic) than "Late" type galaxies (Spiral). This conforms with the fact which says "Elliptic galaxies tend to cluster more strongly than any other galaxy objects" (Gray et al., 2002).

3.4.2.2 Cliques Cardinalities

Figure 3.7 shows the clique cardinalities in the "Main" galaxies. It shows that cliques with cardinality between 2 and 5 (small cliques) are more frequent than large cliques. In other words, in the universe there are no large clusters of galaxies. We mean by large clusters, large number of galaxy objects that are in the same neighborhood of each other. Although in this experiment we used very large threshold (4 Mpc), but we obtained a large number of small cliques.

3.4.2.3 GridClique Performance

Since the previously proposed algorithms, which enumerate maximal clique patterns, are not specifically designed to mine the SDSS, a Naïve algorithm was implemented on the basis of brute force approach to obtain bench marks – this allows us to check the completeness of our algorithm. In this section we show the effect of two factors on the GridClique algorithm, namely, distance and number of spatial objects. This section gives

---

[5]The process of categorizing the galaxies objects is provided in Appendix A.

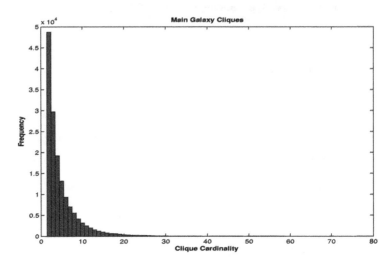

Figure 3.7: Cliques cardinalities for Main galaxies using threshold = 4 Mpc.

a comparison between the GridClique and Naïve algorithms as well.

Figure 3.8 shows the runtime of the GridClique algorithm with various numbers of objects (galaxies) and distance values. It illustrates that the runtime increases slightly as the number of objects and distance increase. The distance and the number of objects were changed in increments of 1 Mpc and 50K, respectively.

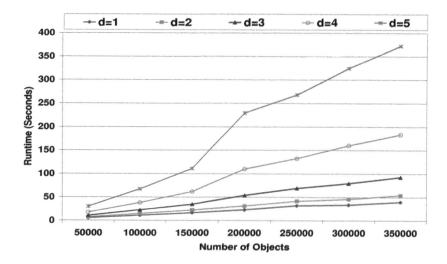

Figure 3.8: The runtime of GridClique using different distances and number of objects. The distance and the number of objects were changed in increments of 1 Mpc and 50K, respectively.

To explain further, when the distance increases the grid size increases. By increasing number of objects at the same time, it allows more objects to appear in the same cells or in the neighbor cells of the grid. In other words, increasing the number of objects increases the cell density. Hence, the two factors (distance and number of objects) affect the runtime of the GridClique algorithm.

In Figure 3.9 we show the performance of the GridClique algorithm using large distances. We changed the distance and the number of objects in increments of 5 Mpc and 5K, respectively. This experiment shows that although the distances is large, the GridClique algorithm run time trends are similar to those when the distance is small, because the sparse nature of the astronomy data.

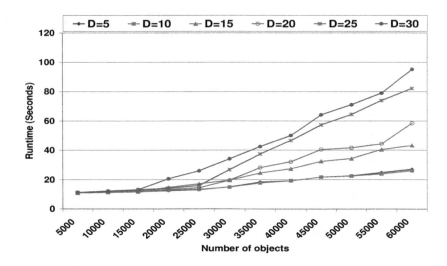

Figure 3.9: The runtime of GridClique using different large distances and small number of objects. The distance and the number of objects were changed in increments of 5 Mpc and 5K, respectively.

Figure 3.10 shows the effects of two factors (number of objects and the distance) on the Naïve algorithm runtime. It is clear that the algorithm is not affected when using different distances. However, its runtime increases exponentially as the number of objects increases.

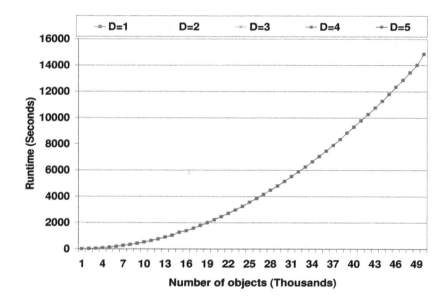

Figure 3.10: The runtime of the Naïve algorithm.

We have carried out an experiment to compare the GridClique performance with the Naïve. Figure 3.11 shows that GridClique outperforms the Naïve algorithm with a difference of several order of magnitudes! We have used a distance of 1Mpc; the number of objects was changed in increments of one thousand objects. The run time is given in logarithmic scale.

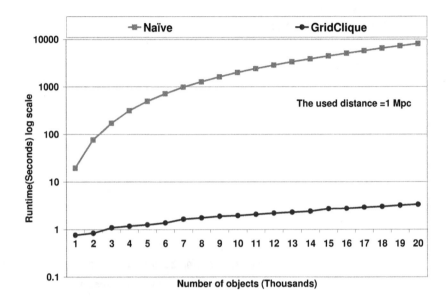

Figure 3.11: Runtime comparison between the GridClique and the Naïve algorithms. The used distance was 1 Mpc.

3.4.2.4   Association Rules Mining Performance

Since most co-location mining algorithms are based on the Apriori algorithm, we use this as the comparison. That is, we evaluate both GLIMIT and Apriori for the interesting pattern mining task of Figure 3.3.

Figure 3.12 shows the number of interesting patterns found on the different sets of cliques. The experiment was carried out for the three type of relationships (Non-Complex relationship, Complex w/o Negative relationship, and Complex w Negative relationship). The MinPI threshold was changed in increments of 0.05.

Figures 3.13, 3.14 and 3.15 show the runtime of the pattern mining. It is clear that

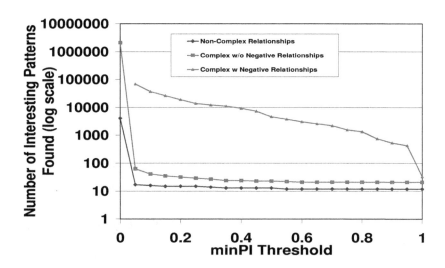

Figure 3.12: Number of interesting patterns found.

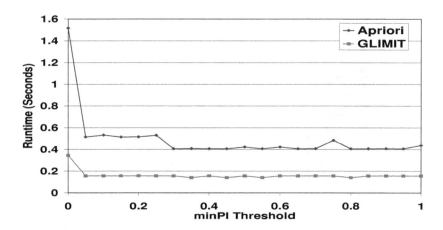

Figure 3.13: Runtime on non-complex maximal cliques. The MinPI threshold was changed in increments of 0.05.

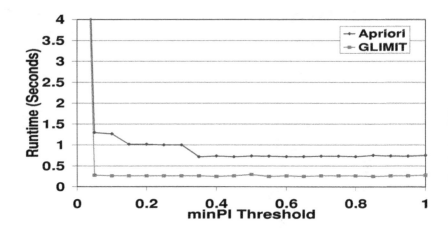

Figure 3.14: Runtime on complex maximal cliques without negative patterns. The MinPI threshold was changed in increments of 0.05.

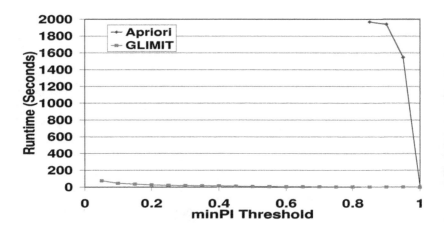

Figure 3.15: Runtime on complex maximal cliques with negative patterns. The MinPI threshold was changed in increments of 0.05. We set an upper limit of 2,000 seconds (33 minutes).

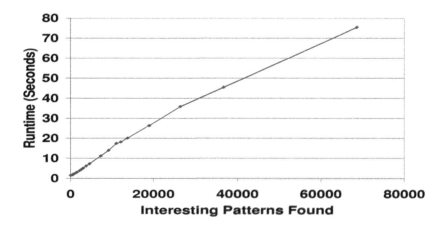

Figure 3.16: The runtime of GLIMIT on complex maximal cliques with negative patterns, versus the number of interesting patterns found. The MinPI threshold was changed in increments of 0.05.

GLIMIT easily outperforms the Apriori technique. In particular, we draw the readers attention to the difference between the run-times when negative items are involved; namely, Figure 3.15. **For example, with a** $minPI$ **threshold of** 0.85, **Apriori takes** 33 *minutes* (1967 **seconds), while GLIMIT takes only** 2 *seconds*. **This is a difference of almost three orders of magnitude!**

As can be seen from Table 3.5, the use of negative types increases the average transaction width substantially. This has a large influence to the runtime of the Apriori algorithm, due to the support counting step where all subsets (of a particular size) of a transaction must be generated. This is not true of GLIMIT, which runs in roughly linear time in the number of interesting patterns found, as can be seen in Figure 3.16. The "non-complex" and "complex w/o negative" datasets, due to their small average transaction width, may be considered easy. The "complex w negative" dataset is very difficult for Apriori, but very easy for GLIMIT. Indeed, even with a $minPI$ threshold of 0.05 it takes only 76 seconds to mine 68, 633 patterns.

Rule Number	Neighborhood distance = 1 Mpc Minimum Confidence = 0.75
1	LrgEarly → LrgLate
2	LrgLate → LrgEarly
3	LrgEarly+ → −LrgLate
4	LrgLate+ → −LrgEarly
5	MainEarly+ → −MainLate
6	MainLate+ → −MainEarly

Table 3.6: Sample of association rules produced by our MCCR technique, where the antecedents and consequents of these rules are galaxy-type objects.

3.4.2.5 Interesting Rules from SDSS

By applying the *itemset mining* algorithm on the maximal cliques which are generated by GridClique, very interesting rules have been discovered. Table 3.6 lists sample of these interesting rules. Rules 1 and 2 show that while single instances of "Early" and "Late" galaxies are found in the neighborhood of each other, their clusters are found in isolation.

Rules 3 and 5 answer the question mentioned in Figure 3.2. That means, the existence of a cluster of elliptical galaxies "Early" repels the existence of spiral "Late" ones.

3.5 Summary and Conclusions

In this chapter, we demonstrated the problem of mining complex co-location rules (MC-CRs). That is, a systematic approach to mine complex spatial co-location pattern in Sloan Digital Sky Survey (SDSS) data. We defined the term *maximal clique* in the context of mining complex spatial co-location. Maximal clique is fundamental to our work.

The MCCRs approach consists of two mining steps: First, it enumerates efficiently maximal clique patterns. In order to achieve the first mining step, we have proposed a heuristic (GridClique) based on a divide-and-conquer strategy that considers all spatial objects as points in a plane. Then it divides the plane into grid structure which helps in reducing the search space. The reported maximal cliques are considered to be the transactional data. MCCRs then uses the transactional data for mining interesting co-location rules using an association rule mining technique.

Most work in this area has used an Apriori style algorithm to do the association rule mining step. We showed that GLIMIT is a much better choice, especially when complex patterns are involved. We argued that complex patterns only make sense in the context of maximal cliques. Using maximal cliques allowed us to easily split the clique generation from the interesting pattern mining tasks and avoid redundant cliques.

The achieved results conformed to real facts in the astronomy domain and this has weighed up our proposed method favorably.

Chapter 4

Mining Complex Spatio-Temporal Patterns

This chapter proposes a method to tackle the curse of dimensionality in large spatio-temporal datasets. It is organized as follows: We provide an introduction in Section 4.1. Then in Section 4.2, we give an overview of the related research on movement patterns. The approximation algorithm is presented and analyzed in Section 4.3. In Section 4.4, our proposed approach of combining random projections with approximation algorithms to efficiently process long-duration ST queries will be outlined. In Section 4.5, the design of the experiments and the results are discussed. Section 4.6 gives a comparison between random projections (RP) and principle components analysis (PCA). Table 4.1 lists the notations used in this chapter[1].

---

[1] This chapter is based on the following publications:

- Ghazi Al-Naymat, Sanjay Chawla and Joachim Gudmundsson. **Random Projection for Mining Long Duration Flock Pattern in Spatio-Temporal Datasets**. In communication with the GeoInformatica, 2008 (Al-Naymat et al., 2008a).

- Ghazi Al-Naymat, Sanjay Chawla and Joachim Gudmundsson. **Dimensionality Reduction for Long Duration and Complex Spatio-Temporal Queries.** The 2007 ACM Symposium on Applied Computing (ACM SAC). Seoul, Korea. March 11–15, 2007. Pages (393–397) (Al-Naymat et al., 2007).

Symbol	Description
GPS	Global Positioning System
ST	Spatio-Temporal
$\mathbb{R}$	Real numbers
$\mathbb{N}$	Natural numbers
PCA	Principle Components Analysis
RP	Random Projections
REMO	RElative MOtion
d	Number of dimensions
n	Number of data points, such as rows, objects
r	Radius
τ	Number of time-stamps
κ	Number of desired dimensions
R	Random matrix
r_{ij}	An entry of a matrix

Table 4.1: Description of the notations used.

4.1 Introduction

The most common type of spatio-temporal (ST) data consists of movement traces of point objects. The widespread availability of GPS enabled mobile devices and location-aware sensors have led to an explosion of generation and availability of this type of ST data. GPS devices promote the capture of detailed movement trajectories of people, animals, vehicles and other moving objects that open new options for a better understanding of the processes involved. This has lead to the analysis of moving objects in a multitude of application areas, such as socio-economic geography (Frank, 2001), transport analysis (Qu et al., 1998), animal behavior (Dumont et al., 2005) and in defense and surveillance areas (Ng, 2001).

The number of projects tracking animals has in recent years increased rapidly (wtp, 2008). An example of a project which is continuously generating ST data is related to the tracking of caribou in Northern Canada. Since 1993 the movement of caribou is being tracked through the use of GPS collars with the underlying expectation that the data collected will help scientist understand the migration patterns of caribou and help them locate their breeding and calving locations (pch, 2007). While the number of caribou tagged at a given time is small, the length of the temporal data associated with each caribou is long. As we will see one of the major challenges in ST query processing and patten discovery is to efficiently handle "long-duration" ST data. Very interesting ST-queries can be formulated on this particular dataset. For example, *how are herds formed and does herd membership change over time? Are there specific regions in Northern Canada where caribou herds tend to rendezvous?*

Another example is the tracking of vehicles. The collection and use of GPS tracker data offers a solution for producing detailed and accurate journey time and journey speed outputs. As the number of vehicles that contains GPS devices grows, the collected data is a prosperous source of traffic information that facilitates transport planning (Storey and Holtom, 2003). The recent Swedish Intelligent Speed Adaptation (ISA) study (Wolf et al., 2004) included the installation of GPS devices in hundreds of cars in three Swedish medium size cities; Borlänge, Lund, and Lidköping, where the vehicles were traced for up to 2 years. This dataset contains a wealth of travel behavior information. To analyze this data requires an automated process that can collect travel behavior details from the collected GPS trajectories.

A more sombre example project involving ST data is related to the study of "pandemic preparedness". How does a contagious disease get transmitted in a large city given the movement of people across the city? A recent workshop held on spatial data mining (pan, 2006) exclusively focused on how to use spatial and ST data mining techniques to help answer such questions. The ST-queries that we will discuss can readily be applied to better understand this scenario.

The aim of this research is to develop general tools to compute simple movement patterns efficiently, hence, our focus is to find a simple, and general, but useful definition of a movement pattern while keeping in mind the possibility of computing it. This general approach has already found applications in the area of defence and surveillance where analysts use these tools to obtain patterns that might be of interest (i.e., as a preprocessing step to detect movements that should be analyzed further).

In practice the definition of flock should be customized to the application domain. For example, a caribou flock should certainly include both geographical features, such as inter visibility between caribou, and expert knowledge about the behavior of caribou.

To abstract the problem we assume that we are given a set P of n moving point objects $p_1, \ldots, p_n$ whose locations are known at τ consecutive time steps $t_1, \ldots, t_\tau$. By connecting every pair of consecutive locations by a straight-line segment the trajectory of each object is a polygonal line that can self-intersect. For brevity, we will call moving point objects *entities* from now on.

As is standard in the literature in this area (Benkert et al., 2006; Gudmundsson et al., 2008; Laube and Imfeld, 2002; Mamoulis et al., 2004) we will assume that the velocity of an entity along a line segment of the trajectory is constant. At first glance this looks as a major constraint since there are many more reasonable models (Tremblay et al., 2006), e.g. Bézier, hermite and cubic splines. However, the main proof only requires the following property: if two entities are close to each other at two consecutive time steps then they are assumed to be close to each other in between the two time steps. We believe that this is a reasonable assumption.

4.1.1   Main Contribution and Scope

The most well-known algorithms for computing fixed subset flocks either have a quadratic dependency on n (Gudmundsson and van Kreveld, 2006) (number of entities) or an exponential dependency on the duration of the flock (Benkert et al., 2006). This became the motivation of this work, and our main contributions are:

- We can use random projection to manage the exponential dependency while provably retaining a high-quality solution.

- We will present experimental results which will confirm that using random projection, as a preprocessing step, can effectively help overcome the "curse of dimensionality" for ST pattern processing.

- We specifically focus on using random projection as a dimensionality reduction technique to retrieve long-duration flock patterns.

- We discuss the reasons behind choosing random projection as the technique to be used in this type of problems.

4.2 Related Work

The problem of detecting movement patterns in spatio-temporal data has recently received considerable attention from several research communities, e.g., geographic information science (Gudmundsson et al., 2008; Shirabe, 2006), data mining (du Mouza and Rigaux, 2005; Jeung et al., 2008; Kollios et al., 2001; Koubarakis et al., 2003; Verhein and Chawla, 2008), data bases (Asakura and Hato, 2004; Güting et al., 2003, 2000; Güting and Schneider, 2005; Park et al., 2003; Wolfson and Mena, 2004; Wolfson et al., 1998), and algorithms (Buchin et al., 2008a,b; Gudmundsson et al., 2007). One of the first movement pattern studied (Jensen et al., 2007; Jeung et al., 2007, 2008; Kalnis et al., 2005; Laube and Imfeld, 2002) was moving clusters. It is not surprising that this was one of the first patterns to be studied since it is the ST equivalent of a point cluster in a spatial setting. A moving cluster in a time interval T consists of at least m entities such that for every point in time within T there is a cluster of m entities. The set of entities might be much larger than m, thus entities may join and leave a cluster during the cluster's life-time. A moving cluster is also called a *variable subset flock*. Closely related to moving clusters is the *flock* pattern, or *fixed subset flock* (Gudmundsson and van Kreveld, 2006). This problem has been studied in several papers (Benkert et al., 2006; Gudmundsson and van Kreveld, 2006; Gudmundsson et al., 2007; Jeung et al., 2007; Laube and Imfeld, 2002; Laube et al., 2004). Even though different papers use slightly different definitions the main idea is that a flock consists of a fixed set of entities moving together as a cluster during the duration of the flock pattern.

More recently, further movement patterns have been studied. Jeung et al. (2008) modi-fied the definition of a flock to what they call a convoy, where a group of entities forms a convoy if they are density-connected. Intuitively, two entities in a group are density-connected if a sequence of objects exists that connects the two objects and the distance be-tween consecutive objects does not exceed a given constant. Gudmundsson et al. (2008) developed approaches to detect leadership patterns. They proposed a definition of a pat-tern to be the geometrical relation of one individual moving in front of its followers such that all the followers can 'see' the leader. Similar results were recently obtained for "single file" patterns (Buchin et al., 2008a).

In this chapter we will focus our attention on the fixed subset flock pattern. Laube and Imfeld (2002) proposed the REMO framework (RElative MOtion) which defines similar behavior within groups of entities. They define a collection of ST patterns on the ba-sis of similar direction of motion or change of direction. Laube et al. (2004) extended the framework by not only including direction of motion, but the location itself. They defined several ST patterns, including flock (co-ordinately moving close together), lead-ership (spatially leading a move of others), convergence (converging towards a spot), and encounter (gathering at a spot), and gave algorithms to compute them efficiently.

However, these algorithms only consider each time step separately, that is, given $m \in \mathbb{N}$

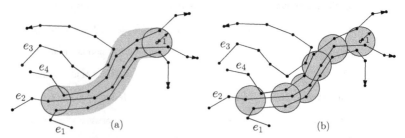

Figure 4.1: Illustrating a flock pattern among four trajectories. If $m = 3$ and the radius $r = 1$ then the longest duration flock lasts for six time steps. Entities e_1, e_2 and e_4 form a flock since they move together within a disk of radius r, while e_3 is excluded in this pattern. Figure(a) illustrates a flock using Definition 4.1 while Figure (b) illustrates a flock using Definition 4.2.

and $r > 0$ a flock is defined by at least m entities within a circular region of radius r and moving in the same direction at a given point in time. Benkert et al. (2006) argued that this is not enough for many practical applications, e.g., a group of animals may need to stay together for days or even weeks before it is defined as a flock. They proposed the following definition of a flock that takes the minimum duration (k) into account (Figure 4.1); m, k and r are given as parameters, and are hence fixed values.

Definition 4.1 *(m, k, r)-flock$_A$ - Given a set of n trajectories where each trajectory consists of τ line segments, a flock in a time interval $I = [t_i, t_j]$, where $j - i + 1 \geq k$, consists of at least m entities such that for every point in time within I there is a disk of radius r that contains all the m entities, and $m, k \in \mathbb{N}$ and $r > 0$ are given constants.*

Benkert et al. (2006) proved that there is an alternative, and algorithmically simpler, definition of a flock that is equivalent provided that the entity moves with a constant velocity along the straight line between two consecutive points. As discussed in the introduction their proof does not require this property, it only requires that if two entities are close to each other at two consecutive time steps then they are assumed to be close to each other in between the two time steps. We believe that this is a reasonable assumption.

If this is true then the following definition is equivalent to the above (Figure 4.1b).

Definition 4.2 (m, k, r)-*flock$_B$ - Given a set of n trajectories where each trajectory consists of τ line segments a flock in a time interval $[t_i, t_j]$, where $j - i + 1 \geq k$ consists of at least m entities such that for every discrete time step t_ℓ, $i \leq \ell \leq j$, there is a disk of radius r that contains all the m entities.*

The latter definition only discrete time points are considered, while the former definition considers time in between two consecutive time steps, see Figure 4.1.

Definition 4.3 *A flock query (m, k, r) returns all (m, k, r)-flock$_B$.*

In the remainder of this chapter we refer to Definition 4.2 whenever we discuss flocks. Using this model, Gudmundsson and van Kreveld (2006) showed that computing the longest duration flock and the largest subset flock is NP-hard to approximate within a factor of $\tau^{1-\varepsilon}$ and $n^{1-\varepsilon}$, respectively, for any $\varepsilon > 0$.

Benkert et al. (2006) described an efficient approximation algorithms for reporting and detecting flocks, where they let the size of the region deviate slightly from what is specified. Approximating the size of the circular region within a factor of $\Delta > 1$ means that a disk with radius between r and Δr that contains at least m objects may or may not be reported as a flock while a region with a radius of at most r that contains at least m entities will always be reported. Their main approach is a $(2 + \varepsilon)$-approximation with running time $T(n) = O(\tau n k^2 (\log n + 1/\varepsilon^{2k-1}))$. Even though the dependency on the number of entities is small the dependency on the duration of the flock pattern is exponential. Thus, the main remaining open problem in (Benkert et al., 2006) is to develop a robust algorithm with a smaller dependency on k, which is exactly the focus of this chapter.

4.3 Approximating Flock Patterns

The nature of a flock is that they involve a "sufficiently large" group of entities passing through a "small" area. This is formalized by the parameters m and r in the definition that represent the number of entities and radius of the region, respectively. An exact value of m and r has no special significance - 20 caribou meeting in a circle of radius 50 is as interesting as 19 caribou meeting in a circle of radius 51. Therefore the use of approximation algorithms is ideally suited for these scenarios (Gudmundsson et al., 2007).

Our approach can be seen as an extension, or modification, of the approach presented by Benkert et al. (2006). We will therefore briefly describe their approach in this section.

4.3.1 Previous Approach

The input is a set P of n trajectories, where a trajectory p_γ is generated by a moving entity e_γ, $1 \leq \gamma \leq n$. Each trajectory p_γ is a sequence of τ coordinates in the plane $(x_1^\gamma, y_1^\gamma), (x_2^\gamma, y_2^\gamma), \ldots, (x_\tau^\gamma, y_\tau^\gamma)$, where (x_j^γ, y_j^γ) is the position of entity e_γ at time t_j.

The basic idea builds on the fact that a polygonal line with k vertices in the plane can be modeled as a point in $2k$ dimensions. The trajectory of an entity p in the time interval $[t_i, t_j]$ is described by the polygonal line $p(i, j) = \langle (x_i, y_i), (x_{i+1}, y_{i+1}), \ldots, (x_j, y_j) \rangle$, which corresponds to a point $p'(i, j) = (x_i, y_i, x_{i+1}, y_{i+1}, \ldots, x_j, y_j)$ in $2(j - i + 1)$-dimensional space.

The first step when checking whether there is a flock in the time interval $[t_i, t_{i+k-1}]$ is to map each trajectory into a point in $\mathbb{R}^{2k}$. The second step is to characterize a flock in high-dimensional space. It turns out that a flock can be expressed in terms of a high-dimensional *pipe*.

Definition 4.4 *((Benkert et al., 2006))*

An (x, y, i, r)-pipe in $\mathbb{R}^{2k}$ is the region: $\left\{ (x_1, \ldots, x_{2k}) \in \mathbb{R}^{2k} \mid (x_i - x)^2 + (x_{i+1} - y)^2 \leq r^2 \right\}$.

An (x, y, i, r)-*pipe* is an unbounded region in $\mathbb{R}^{2k}$ and contains all the points that are only restricted in two of the $2k$ dimensions (namely in dimensions i and $i + 1$) and when projected on those two dimensions lie in a circle of radius r around the point (x, y). Equivalence 4.1 gives the key characterization of flocks.

Equivalence 4.1 *(Equivalence 1 in (Benkert et al., 2006))*

Let $F = \{p_1, \ldots, p_m\}$ be a set of trajectories and let $I = [t_1, t_k]$ be a time interval. Let $\{p_1', \ldots, p_m'\}$ be the mappings of F to $\mathbb{R}^{2k}$ w.r.t. I. It holds that:

$$F \text{ is a } (m, k, r)\text{-flock} \quad \Longleftrightarrow$$

$$\exists x_1, y_1, \ldots, x_k, y_k : \forall p \in F : p' \in \bigcap_{i=1}^{k} (x_i, y_i, 2i - 1, r)\text{-pipe.}$$

Equivalence 4.1 just restates the definition of a flock for trajectories as a definition using the corresponding points in $2k$ dimensions. It simply says that if a set of entities form a flock in the time interval $I = [t_1, t_k]$ then the projection of the points corresponding to the trajectories of these entities in the plane spanned by the x_i-axis and y_i-axis, $0 \leq i \leq k$, will lie in a disk of radius r for every two consecutive dimensions.

Equivalence 4.1 has been used by Benkert et al. (2006) to design a Δ-approximation algorithm to find flocks ($\Delta > 1$) by performing a set of n range counting queries (Eppstein et al., 2005) in the transformed space.

Definition 4.5 *((Benkert et al., 2006))*

A Δ-approximation algorithm will report every (m, k, r)-flock, it may or may not report an $(m, k, \Delta r)$-flock and it will not report a (m, k, r')-flock where $r' > \Delta r$.

For each time interval $I = [t_i, t_{i+k-1}]$, where $1 \leq i \leq \tau - k + 1$, do the following computations. For each trajectory p let p' denote the mapping of p to $\mathbb{R}^{2k}$ with respect to I. For each point p' perform a range counting query where the query range $Q(p')$ is the intersection of the k pipes $(x_i, y_i, 2i - 1, 2r)$ and (x_i, y_i) is the position of entity p at time step t_i. Every counting query containing at least m entities corresponds to an $(m, k, 2r)$-flock according to (Benkert et al., 2006), where the same flock may be reported several times.

Since the theoretical bounds for answering range counting queries in high-dimensions are close to linear Benkert et al. (2006) instead used $(1 + \delta)$-approximate range counting queries, to obtain the following result.

Fact 4.1 *(Lemma 7 in (Benkert et al., 2006))*
The algorithm is a $(2+\delta)$-approximation algorithm and requires $O(kn(2^k \log n + k^2/\delta^{2k-1}))$ time and $O(\tau n)$ space, for any $\delta > 0$.

In (Benkert et al., 2006) the skip-quadtree by Eppstein et al. (2005) was used to achieve the theoretical bounds. However, in practice it turns out that a standard quadtree performs slightly better. Let $S = p_1, p_2, \ldots, p_n$ be a set of n points in the plane contained in a square C of length l. A quadtree QT for S is recursively constructed as follows: The root of QT corresponds to the square C. The root has four children corresponding to the four squares of C of length $\frac{1}{2}$. The leaves of QT are the nodes whose corresponding square contains exactly one point. Using a compressed quadtree (Arya et al., 1998) for QT reduces its size to $O(n)$ by removing nodes not containing any points of S and eliminating nodes having only one child. A compressed quadtree for a set of n points in the plane can be constructed in $O(n \log n)$ time, see the book by Samet (2006) for more details.

In (Benkert et al., 2006), it was shown that this approach is very effective for small values of k, they performed experiments with k in the range of 4 to 16. However, since the running time has an exponential dependency on k it is obvious that it cannot handle long-duration flocks very well. As mentioned before this is a serious drawback since flocks may stay together for a very long time.

4.4   Random Projections

Several techniques are available for carrying out the dimensionality reduction, e.g., PCA (Principal Components Analysis), discrete wavelet transform, discrete cosine transform, and random projection (RP) (Achlioptas, 2003). We used in this book the PCA and RP techniques. A wide range of applications that used both techniques were discussed in Chapter 2.

We have chosen to use random projections technique because:

1. Random Projections yield a point-wise bound on the distortion on the distance between two points before and after projection (Theorem 4.1).

2. PCA on the other does not provide any guarantee on point-wise distortion. PCA can be used to project an $m \times n$ matrix A into a rank $\kappa < min(m, n)$ matrix A_κ which minimizes the Frobenius norm[2] $||A - C||_F$ over all matrices of rank κ. Since our objective is to project points into a lower dimension space in order to efficiently use spatial data structures and preserve pair-wise distances, PCA may not be particularly useful.

3. Random Projections are far more efficient than PCA. For Random Projections the time complexity of mapping $mn-$dimensional points into κ dimensions is $O(Nd\kappa)$. On the other time complexity of PCA is $O(mn^3)$.

---

[2]It is also known as the Euclidean norm.

Thus in order to handle long-duration flocks, we extend the algorithm of finding flocks by adding a preprocessing step. The added step is intended to reduce the number of dimensions so that we can apply the algorithm in (Benkert et al., 2006). However, a flock is defined as a group of entities that at all times lie within a disk in the plane of radius r. If two entities lie within the same query range that means that the distance between them is at most $2r$ in every dimension. However, the distance between them in $2k$-dimensions may be roughly $r \cdot \sqrt{2k}$ (Definition 4.4). This observation makes it clear that we cannot perform a dimensional reduction and then use the same approach as above since the error will be too large.

However, if we slightly modify the definition, the generalization can still be performed.

Definition 4.6 *(m, k, r)-flock$_C$ - Given a set of n trajectories where each trajectory consists of τ line segments a flock f in a time interval $[t_i, t_j]$, where $j - i + 1 \geq k$ consists of at least m entities such that for every pair of trajectories $p, q \in f$ it holds that $\sum_{\ell=i}^{j} |p_\ell - q_\ell| \leq r \cdot \sqrt{2k}$.*

That is, instead of using a maximum distance of r in each time step we bound the sum of the differences. Intuitively this means that two entities that are very close to each other in all but one time step may still belong to the same flock in the new definition while this would not be possible in the definition by Benkert et al. (2006).

Recall that in a step we consider a set of n points in $\mathbb{R}^{2k}$. Our first approach reduces the dimensions by using random projections by (Johnson and Lindenstrauss, 1982), (Achlioptas, 2003) and (Indyk and Motwani, 1998).

Theorem 4.1 *(Achlioptas, 2003)*

Let P be an arbitrary set of n points in $\mathbb{R}^d$, represented as an $n \times d$ matrix A. Given

$\beta, \epsilon > 0$ let

$$\kappa_0 = \frac{4 + 2\beta}{\epsilon^2/2 - \epsilon^3/3} \log n.$$

For integer $\kappa \geq \kappa_0$, let R be a $d \times \kappa$ random matrix with $R(i, j) = r_{ij}$, where r_{ij} are independent random variables from the following probability distributions:

$$r_{ij} = \sqrt{3} \times \begin{cases} +1 & \text{with probability} & 1/6 \\ 0 & .. & 2/3 \\ -1 & .. & 1/6. \end{cases}$$

Let $E = \frac{1}{\sqrt{\kappa}} \cdot AR$ and let $f : \mathbb{R}^d \to \mathbb{R}^\kappa$ map the ith row of A to the ith row of E. We have, for all $u, v \in P$ it holds that:

$$(1 - \epsilon)\|u - v\|^2 \leq \|f(u) - f(v)\|^2 \leq (1 + \epsilon)\|u - v\|^2,$$

with probability at least $1 - n^{-\beta}$.

Recall that all the trajectories with k vertices in the plane can be modeled as a point in $2k$ dimensions. Thus after the transformation we have a set P of n points in $2k$ dimensions. Next, instead of performing n range counting queries, as in (Benkert et al., 2006), we first apply the random projection on P to obtain a set P' of n points in $\kappa = \frac{4+2\beta}{\epsilon^2/2-\epsilon^3/3} \log n$ dimensions. Then, for each entity p perform a $(1 + \delta)$-approximate range counting query where the query range $Q(p')$ is the κ-dimensional ball of radius $O(\sqrt{k})$ and center at p' which is the mapping of p in P'.

4.4.1 A Theoretical Analysis

Here we briefly give a probabilistic analysis on how the dimensional reduction affects the algorithm. We say that an algorithm is an α-probabilistic $(m, k, \Delta r)$-approximation algorithm if all (m, k, r)-flock are reported with probability α, an $(m, k, \Delta r)$-flock may or may not be reported, while no $(m, k, \Delta' r)$-flock will be reported with probability α.

Theorem 4.2 *The modified algorithm is a $(1 - n^{1-\beta})$-probabilistic $(m, k, \Delta r)$-approximation algorithm with running time $O(\tau n k^2 (\log n + 1/\delta^{2\kappa}))$ time, where $\kappa = \frac{4+2\beta}{\varepsilon^2/3 - \varepsilon^3/3} \log n$ and $\Delta = 2(1 + \delta) \cdot (1 + \varepsilon)^2$ for any constants $\varepsilon, \delta > 0$ and $\beta > 1$.*

Proof Consider the trajectories and let $P = \{p_1, \dots, p_n\}$ be the corresponding points in $2k$-dimensions, and let $f(p_i)$ be the point in P' corresponding to $p_i \in P$.

First we show that each (m, k, r)-flock$_c$ f is reported by the algorithm with high probability. Let p_f be a point in P corresponding to an arbitrary entity of f and assume that f is a flock in the time interval $I = [t_i, t_{i+k-1}]$. We will prove that the approximation algorithm returns an $(m, k, \Delta r)$-flock g such that $f \subseteq g$ with high probability and $\Delta = (1 + \delta) \cdot (1 + \varepsilon)^2$.

According to Definition 4.6 there exists a $2k$-dimensional ball with radius $r \cdot \sqrt{2k}$ that contains the points in P corresponding to the entities in f. According to Theorem 1 the corresponding points in P' will lie within a κ-dimensional ball of radius $(1 + \varepsilon) r \cdot \sqrt{2k}$ with probability at least $(1 - n^{1-\beta})$, which is the probability that a flock g containing the entities in f is reported.

For the stated bound to hold we have to prove that the probability that an $(m, k, \Delta' r)$-flock$_c$ is reported is small for $\Delta' > \Delta$. Let g be a flock reported by the algorithm in the interval I. The points in P' corresponding to the entities in g must lie within a κ-dimensional ball of radius $(1 + \delta) \cdot ((1 + \varepsilon) r \cdot \sqrt{2k})$ and center at a point $f(p') \in P'$. Recall that the algorithm performs $(1 + \delta)$-approximate range counting queries. Using Theorem 4.1 it then follows that the maximum distance between p' and every point corresponding to an entity in g is at most $(1 + \varepsilon)((1 + \delta) \cdot ((1 + \varepsilon) r \cdot \sqrt{2k})) = (1+\delta) \cdot ((1+\varepsilon)^2 r \cdot \sqrt{2k})$ with probability $(1 - n^{1-\beta})$. From the triangle inequality we get

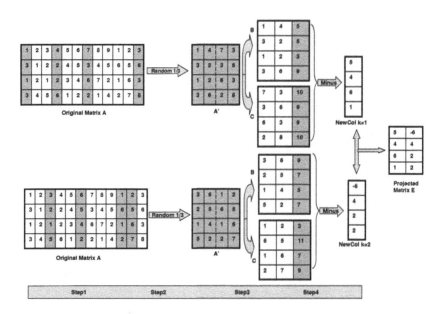

Figure 4.2: Random projection using "DB friendly" operations. To compute the random projection without actually generating the matrix with random entries, project the original database and process these new projections (tables) further to generate the final database that consists of low number of dimensions (i.e., number of columns).

that the distance between any two points in g is bounded by $2(1 + \delta) \cdot ((1 + \varepsilon)^2 r \cdot \sqrt{2k})$ which completes the proof of the theorem. □

4.4.2 Random Projection in a Database Management System

In 2003, (Achlioptas, 2003) showed how a random projection can be carried out using "database friendly" operations, i.e., computing a random projection without actually generating the matrix with random entries. We clarify the process with the help of an example shown in Figure 4.2.

Given an integer κ and a table A of dimension $size(A) = n \times d$, where n is the number of rows, d is the number of columns and κ is the reduced dimension. We want to generate a new table E of size $n \times \kappa$ without actually multiplying A with a randomly generated matrix.

For each i, $1 \leq i \leq \kappa$ we produce a new field in the new table and then at the end we merge them together. Effectively, we repeat for each i the following steps:

1. Uniformly at random select $1/3$ of A's attributes. Call the obtained table A'. In other words, we project A to A' (Figure 4.2):

$$A' = \prod A_{d/3}.$$

 The size of A' is $n \times e$, where e is $1/3$ of the columns in the original table, i.e., $e = \text{round}(d/3)$.

2. Uniformly at random vertically partition A' into two equal column-sized tables B and C:

$$B = \prod_{1}^{e/2} A' \quad \text{and} \quad C = \prod_{e/2+1}^{e} A',$$

 as shown in Figure 4.2.

3. We combine B and C to generate a new column as follows. We first create a $NewCol_1$ from B and $NewCol_2$ from C. The third column in each B and C are subtracted to form the columns in the projected matrix E:

$$NewCol_1 = \left\{ \begin{array}{c} \sum_{j=1}^{e/2} B_{1,j} \\ \vdots \\ \sum_{j=1}^{e/2} B_{n,j} \end{array} \right\} \quad \text{and} \quad NewCol_2 = \left\{ \begin{array}{c} \sum_{j=1}^{e/2} C_{1,j} \\ \vdots \\ \sum_{j=1}^{e/2} C_{n,j}. \end{array} \right\}.$$

4. The ith column of table E is defined as $NewCol_2 - NewCol_1$.

4.5 Experiments, Results and Discussion

We report on the experiments that we have carried out to determine how the use of random projection as a preprocessing step will effect the accuracy of detecting flocks in long-duration ST datasets.

4.5.1 Experimental setup and datasets

All experiments were carried out on a Linux operated PC with a Pentium-4 (3.6 GHz) processor and 2 GB main memory. The data structures and algorithms were implemented in C++ and compiled with the GNU compiler.

We use both real datasets and synthetic datasets to demonstrate the efficiency of our approach. The use of synthetic data allows us to control the number of flocks present in the data, which helps in determining the correctness of our approach experimentally. Appendix B provides details about the ST data preparation process as well as the techniques used to obtain it. A description, however, of the used datasets is given below.

4.5.1.1 Synthetic datasets

Twenty datasets with varying number of points, number of flocks and duration of flocks were created. In particular, five datasets each of size 16K, 20K, 32K, 64K and 100K were seeded with 32, 40, 64, 128 and 200 flocks respectively of duration (number of time step) 8, 16, 500 and 1000. The size of each flock was set to 50 entities and the radius was fixed to 50 units. In the original data (before the random projection), each point coordinate was selected from the integer interval $[0, \ldots, 2^{16}]$.

In our experiments, which use synthetic data, we always searched for flocks with at least 50 entities within a circle of radius 50 and full time duration (8, 16, 500 or 1000).

4.5.1.2 Real world datasets

Three different real world datasets were used in our experiments. These datasets were generated by three different projects, namely, Network Dynamics and Simulation Science Laboratory (NDSSL), Caribou Herd Satellite Collar Project, and Mobile Users profiles. A brief description about these projects is as follows:

- **NDSSL Data:**

   Network Dynamics and Simulation Science Laboratory (NDSSL) has produced several synthetic datasets that have been released to the academic community for research purposes (NDS, 2007). The dataset used in this chapter represents the movement of a synthetic population (derived from the census data) of the city of Portland. It shows a number of activities that people do in daily life, such as working, studying, shopping, recreation, and picking up and dropping off passengers.

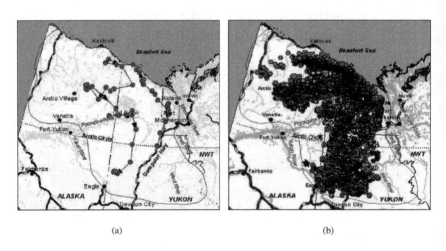

(a)   (b)

Figure 4.3: Caribou's locations in Northern Canada (pch, 2007).

The data is related to the study of "pandemic preparedness". Pandemic influenza viruses have demonstrated their ability to spread worldwide within months, or weeks, and to cause infections in all age groups. Retrieving movement patterns can help in understanding the movement of people in a city and in understanding how pandemic may spread.

- **Caribou Herd Satellite Collar Project:**

  This is a collaboration between a number of wildlife agencies that use satellite radio-collars to track and collect information about the migration of the caribou herds (pch, 2007). At the start of the project, 10 cow caribou were captured in October and November of 1997. Until the end of 2006, the total number of caribou that have been captured was 44. Over that period of time, these cows were tagged with satellite collars. Over time some collars were added and changed as new caribou were tagged, caribou died, or the collars fell off. The total home area that has been used to track the caribou herds is roughly $260,000km^2$, between Kaktovik, Alaska to Aklavik, NWT to Dawson City, Yukon. Figure 4.3 shows the area where caribou cows migrate or meet for calving, this area is bounded by the red line. It shows the problem of visualizing this kind of data. In Figure 4.3(a), one trajectory is shown and in Figure 4.3(b) a subset of all the trajectories is shown in an attempt to visualize the migration behavior.

- **Mobile Users profiles:**

  A realistic simulation approach was developed by Taheri and Zomaya (2005) to simulate mobile user profiles which consist of a collection of different categories of users. These profiles show a variety of users on the basis of their activities. They include, for example, workers, ordinary users, travelers, and users who go out for work only at night. These activities represent the users' movements. By using this generator, an entire population consisting of 10K users has been created.

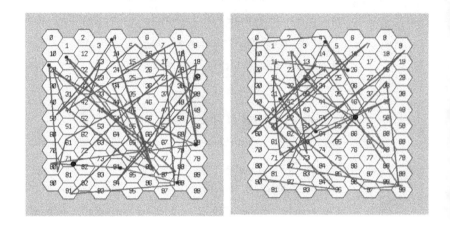

Figure 4.4: Two mobile users' movements for 500 timesteps (Taheri and Zomaya, 2005).

The total number of time steps that has been generated for each user is roughly 500. Figure 4.4 presents the design of the network that was used to represent the mobile users' movements. It consists of 10×10 cells, where each cell represents a circle with a diameter of 60 units. The small dots that appear in the figure show the location when a user makes a phone call. It should be noted that, for the purpose of this work, we do not care about these locations. We focus on the location of the users during their movements regardless of their tasks.

dataset	# Objects	# TimeSteps	# Dimensions	Distance Unit (Radius)	Percentage
NDSSL	10K	18	36	20	0.0007
Caribou	44	359	718	200	2
Mobile users	10K	500	1000	9	0.025

Table 4.2: Datasets description.

The flock radius for each dataset is shown in column 5 in Table 4.2. These values were chosen arbitrarily and will be used as benchmark values. Column 6 contains the relative area of a flock compared to the area of the universe described by each dataset.

The size of each flock was set to at least 3 entities for the Caribou and Mobile users datasets, and to at least 4 entities for the NDSSL data. These numbers were chosen so that the number of flocks in each set is "reasonable".

4.5.2   Random Projection Parameters

The theoretical lower bound for the reduced dimensions using random projection is $\Omega(\frac{ln(n)}{\epsilon^2})$ (Achlioptas, 2003). Notice that the bound is independent of the dimensionality of the original space. For a dataset of size 32K and a distortion of $\epsilon = 0.5$, the dimensionality of the projected space will be at least 165. This is too large for the quadtree indexing structure to handle in practice. However, several research studies (Bingham and Mannila, 2001; Fradkin and Madigan, 2003) have noted that the theoretical bound is quite conservative. For example, Bingham and Mannila (2001) have noted that for *their* image data the theoretical bound required was $\kappa \approx 1600$ but $\kappa \approx 50$ "was enough".

In order to test the difference between the theoretically derived and experimentally acceptable bound (κ) we carried out several experiments for different values on the size of the dataset generated synthetically (n), the dimensionality (d), the error tolerance (ϵ) and the confidence (β). We report on two sets of parameters (Figure 4.5). Notice that while the theoretical bound depends on n (as expected) the experimentally derived bound quickly stabilizes for different values of n. This confirms that the theoretical bound is too conservative and that for practical applications we can use a much smaller κ.

In the end we settled for projecting the data to 32 dimensional space. The primary reason being that the experiments reported in (Benkert et al., 2006) only go up to sixteen time steps (32 dimension) after which the authors note that a range query search using quadtree

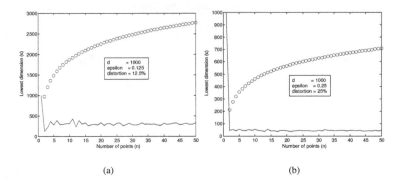

(a)   (b)

Figure 4.5: Illustrating the difference between the theoretical and experimental bounds on the minimum number of dimensions using random projection. The top line is the theoretical bound and the bottom line is derived experimentally using a brute force procedure.

effectively reduces to a linear scan. Since random projection is being used as a pre-processing step, it is reasonable to project to a space where the use of a quadtree is still effective.

Another major difference in the experiments compared to the theoretical results is the choice of the flock radius in the projected space. In Section 4.4, it was shown that the new flock radius should be a factor $\sqrt{2k}$ larger than the original radius; for 500 dimensions that would be approximately a factor of 32. However, it is easily seen that this would only occur in extreme cases so instead we decided to use the original flock radius for the projected data.

4.5.3 Assessment methods

We perform three different assessments of the different approaches. The first one is the performance time in finding the flock patterns, both with and without using random projection as a preprocessing step.

Secondly, we test the accuracy of the approaches by:

1. Comparing the number of reported flocks in the dataset before and after applying the random projection. Optimally we would want the numbers to match.

2. Comparing the retrieved flock's members from each dataset. That is, by comparing the flock's members before and after the projection. A measure *CommonRatio* was used to calculate the ratio of the common members. That is by counting the number of common members between every two flocks (flock before and after projection) and dividing this number by the size of the original flock (Equation 4.1). This ratio is used to find the average accuracy of each data set that is shown in Equation 4.2:

$$CommonRatio = \frac{\#CommonMembers}{s} \quad \text{and} \quad (4.1)$$

$$AvgAcc = \frac{\sum_{i=1}^{n} CommonRatio}{n}, \quad (4.2)$$

where n is the total number of flocks in the dataset, $CommonMembers$ is the number of common members between flocks and s is a positive integer representing the flock's size in the original data.

Thirdly, we considered the interpoint distances before and after the projection. That is, the distance matrix was calculated for all flocks before and after the projection and histograms were plotted to show the difference between the two matrices.

4.5.4   Results

This section demonstrates the experimental results that we achieved after applying random projections on synthetic and real datasets.

	Data		Brute Force(BF)		BF-With Proj		Pipe		Pipe-With Proj	
	Data Size	#TS	#Flocks	Sec	#Flocks	Sec.	#Flocks	Sec.	#Flocks	Sec.
1	16K	8	32	< 1	N.A	N.A	32	< 1	N.A	N.A
2	16K	16	32	< 1	N.A	N.A	32	< 1	N.A	N.A
3	16K	500	32	50	32	< 2	32	(BF) 50+	32	< 2
4	16K	1000	32	59	32	< 2	32	(BF) 59+	32	< 2
5	20K	8	40	< 1	N.A	N.A	40	< 1	N.A	N.A
6	20K	16	40	2	N.A	N.A	40	1	N.A	N.A
7	20K	500	40	82	40	3	40	(BF) 82+	40	2
8	20K	1000	40	87	40	3	40	(BF) 87+	40	2
9	32K	8	64	2	N.A	N.A	64	< 1	N.A	N.A
10	32K	16	64	8	N.A	N.A	64	1	N.A	N.A
11	32K	500	64	206	64	8	64	(BF) 206+	64	3
12	32K	1000	64	216	64	9	64	(BF) 216+	64	3
13	64K	8	128	4	N.A	N.A	128	2	N.A	N.A
14	64K	16	128	16	N.A	N.A	128	3	N.A	N.A
15	64K	500	128	412	128	18	128	(BF) 412+	128	5
16	64K	1000	128	432	128	19	128	(BF) 432+	128	6
17	100K	8	200	12	N.A	N.A	200	4	N.A	N.A
18	100K	16	200	24	N.A	N.A	200	5	N.A	N.A
19	100K	500	200	640	200	27	200	(BF) 640+	200	7
20	100K	1000	200	675	200	28	200	(BF) 675+	200	8

Table 4.3: Summarizing our experimental results, with and without random projection using 16K, 20K, 32K, 64K and 100K data sizes.

4.5.4.1   Synthetic dataset experiments:

The experimental results are shown in Table 4.3 and Figure 4.6. Recall the two basic algorithms described in Section 4.3; the *Brute-Force (BF)* and *Pipe* methods. The BF method does not use any indexing, instead it consists of two nested loops, the outer one specifying a potential flock center and the inner one computing the distance between a point and the potential flock center. If there are at least m points within a ball of radius $2r$ centered at the potential flock center then a flock is reported. In that respect, the BF method is a 2-radius approximation. Notice that the complexity of the BF method is quadratic in the number of entities and only has a small dependency on the number of time stamps. This explains the small increase in time (Table 4.3, rows 3 and 4) for 16K points from 50 seconds to 59 seconds as the number of time steps (TS) increases from

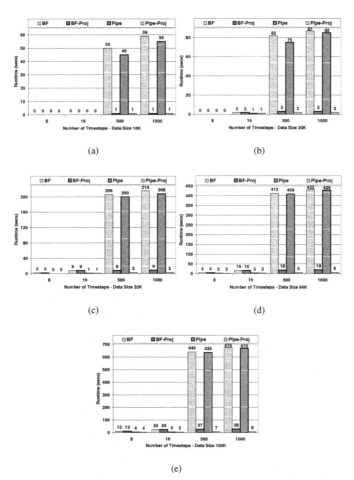

Figure 4.6: Results- with and without random projection using 16K, 20K, 32K, 64K and 100K data sizes.

500 to 1000.

The pipe method is based on Equivalence 4.1 and uses a compressed quadtree as the underlying indexing structure. As expected, the pipe method beats the BF approach for

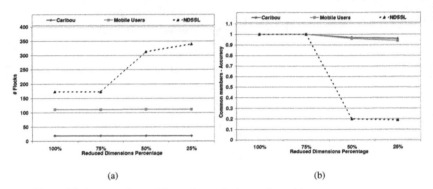

(a)   (b)

Figure 4.7: Accuracy after applying random projections on the real datasets.

low-dimensions. For example, for 32K points, BF requires 2 and 8 seconds to find the
64 flocks for time duration 8 and 16, respectively (rows 9 and 10), while the pipe method
requires less than one second. For higher dimensions (500 and 1000) the quad tree provides no extra advantage, the internal nodes of a quadtree has 2^d children where d is the
number of dimensions. Using 16 time stamps means 32 dimensions which translates to
more than 4 billion quadrants. It is very unlikely that the 32K randomly distributed points
(not in flocks) fall into the same quadrant. This results in a very flat tree and, hence, a
high query time.

We did not apply the random projection preprocessing step for time steps 8 and 16. The
overall cost of random projection is $O(nd\kappa)$ which is around 1 to 3 seconds for the dataset
32K and 100K respectively for the largest time steps.

A key result with the random projection is that we retrieve exactly the same number of
flocks as retrieved without the random projection using the synthetic datasets. Table 4.3
shows that the distortion induced by the random projection is within an acceptable bound
and does not violate the overall correctness.

4.5.4.2 Real dataset results

In the previous section we have shown the performance of our approach as well as the distortion induced by applying the random projection on synthetic datasets. In this section we report on the accuracy of obtaining flocks after applying random projection on real life datasets.

1. *Number of reported flocks before and after the projection.* Table 4.4 presents the total number of reported flocks before and after applying random projection. The original dimensions were reduced by 75%, 50%, and 25%. For the caribou data, exactly the same number of flocks (19 flocks) were discovered for all three projections. For the mobile users data, the exact number of flocks (111 flocks) were discovered for the first two projections (75% and 50%). However, for the third projection (25%) an extra flock was reported (112 flocks), as seen in Figure 4.7(a).

   NDSSL data originally contained a lot of missing locations. Therefore, the data had to be processed before using our method. This was performed in two steps; (a) filling in missing values (locations) and (b) making all trajectories of the same length.

   (a) Missing location in the middle of the trajectory: We assumed that the GPS was faulty and the object location was not sent. Hence the last location was repeated to fill the missing location.

		Before Projection		Projection I (75%)		Projection II (50%)		Projection III (25%)	
Data Name	Flock Size	#TS	#Flocks	#TS	#Flocks	#TS	#Flocks	#TS	#Flocks
NDSSL	≥ 4 Objects	18	173	13	173	9	313	4	400
Caribou	≥ 3 Objects	359	19	269	19	179	19	89	19
Mobile Users	≥ 3 Objects	500	111	375	111	250	111	125	112

Table 4.4: Number of flocks before and after applying random projection onto real life datasets.

(b) If a trajectory is shorter than the others, we assumed that the object stopped before finishing its trip, or the GPS device stopped sending details because of some defects. To handle this we assumed that the object stopped and just repeated the last location until the length of the trajectory was equal to the longest trajectory.

Due to the above assumptions random projection did not work very well when projecting the NDSSL data to very low-dimensions, which clearly can be seen in Table 4.5 and Figure 4.7. For example, if two trajectories were part of the same flock before one of the GPS becomes faulty, will now not be reported as part of the same flock.

2. *Number of common flock members.* After reporting the flock patterns before and after the projection, we compare the retrieved flock members from each dataset. Table 4.5 illustrates the average accuracy for each dataset. For the caribou dataset the accuracy was 100%, 97% and 96% for the three projections 75%, 50% and 25%, respectively. For the mobile users dataset, the accuracy decreased from 100% to 94%. However, for NDSSL dataset, the accuracy deteriorated from 100% to 19% when the number of dimensions is reduced from 75% to 25% of the original dimension. These results are shown in Figure 4.7(b).

	Before Projection	Projection I (75%)		Projection II (50%)		Projection III (25%)	
Data Name	#TS	#TS	Accuracy	#TS	Accuracy	#TS	Accuracy
NDSSL	18	13	100%	9	20%	4	19%
Caribou	359	269	100%	179	97%	89	96%
Mobile Users	500	375	100%	250	96%	125	94%

Table 4.5: Accuracy based on the flock members after the random projections.

The distance matrix was calculated for all flocks before and after the projection and histograms were plotted to show the difference between the two matrices. As expected, the two histograms for each dataset show that the distances in the projected space are smaller than the distances in the original space as noted in Fact 4.1. They show that the two matrices are very similar. Figure 4.8 shows the caribou, the mobile users and the NDSSL flock distributions, before and after applying the random projection 100 times and averaging the results. Table 4.6 shows the number of dimensions before and after applying the random projection.

dataset	Number of original dimensions	Number of dimensions after the projection
NDSSL	36	12
Caribou	718	16
Mobile users	1000	48

Table 4.6: Number of dimensions before and after applying the random projection.

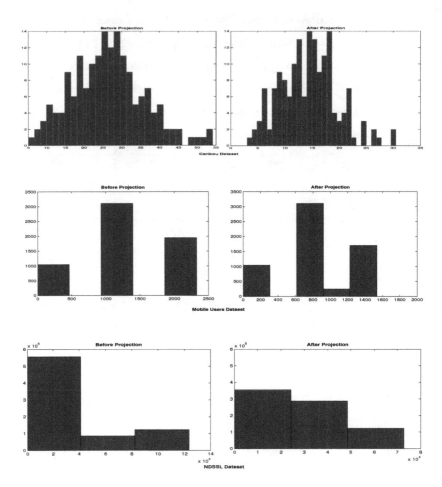

Figure 4.8: Distance matrix distributions for the three datasets flocks (Caribou, Mobile Users, and NDSSL) before and after applying the random projection 100 times and averaging the results.

4.6 Random Projection (RP) and Principal Components Analysis (PCA)

In this section we compare the accuracy of flock discovery between random projection and principal component analysis. We first report a result on a synthetic dataset and then on real datasets described in the previous section. For completeness we begin by giving a brief introduction on Principal Component Analysis.

4.6.1 Principal Components Analysis (PCA)

PCA is a dimensionality reduction technique. It operates by first rotating the coordinate system and then "dropping" some of the dimensions (axis) and effectively projecting the data into a lower dimension space. The new coordinate system is determined based on the direction of the variance in the data. For most real datasets, the majority of the variance is captured in the first few new dimensions and by "dropping" the other dimensions, the "essential structure" of the data is preserved. PCA does not guarantee the preservation of the pairwise distance (even approximately) between data points. Bishop (2006) is a good source for a geometric and algebraic introduction to PCA.

Algebraically, PCA works as follows. Let A be an $m \times n$ matrix, where m is the number of objects each of dimension n. Then A can be decomposed as

$$A = \Sigma_{i=1}^{r} \sigma_i u_i v_i^t.$$

Here, $\sigma_i's$ are the eigenvalues of the mean-centered covariance matrix $A^t A$, $v_i's$ are the corresponding eigenvectors, r is the rank of the matrix A and the $u_i's$ are the new coordinates in space spanned by the new eigenvectors. The decomposition of A is organized in a way such that $\sigma_1 > \sigma_2 > \ldots > \sigma_r$. Geometrically, the $\sigma_i's$ are variance in the direction

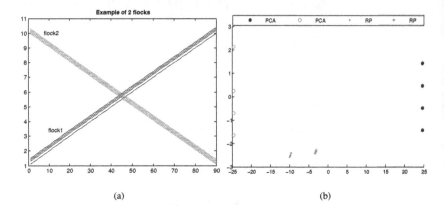

(a) (b)

Figure 4.9: Two flocks before projection (a) and after projection (b). It is clear that RP leads to better flock preservation.

determined by the v_i's. Thus for $k < r$,

$$A = \Sigma_{i=1}^k \sigma_i u_i v_i^t$$

is effectively a lower dimension projection of the objects in A.

4.6.2 Comparison on a Synthetic Dataset

We first compare RP and PCA on a synthetic dataset shown in Figure 4.9(a) which shows two distinct flocks moving along the diagonals of the rectangle. Each flock consists of four objects. We project the dataset on a two-dimensional space using RP and PCA. The results of RP are the average of ten runs of the random matrix. It is clear that RP clusters the elements of the flocks in clean clusters which can be discovered by the flock algorithm in the two-dimensional space. PCA separates the elements of the flock but the inter-element distance within flocks is large.

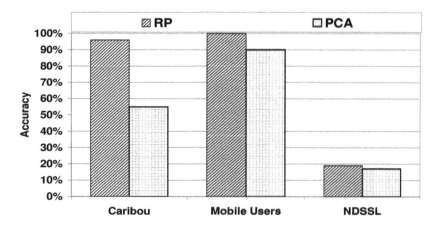

Figure 4.10: Accuracy after applying random projections (RP) and principal components analysis (PCA) on the real datasets.

4.6.3 Comparison on Real Datasets

Again, RP and PCA were applied on the real datasets described in the previous section. The accuracy, as measured by Equation 4.1 and 4.2, after projection onto the two-dimensional dataset are shown in Figure 4.10. Again, RP achieves better accuracy than PCA on all the datasets. It is important to reiterate that RP scales linearly in the number of dimensions while PCA has a cubic dependency. Figure 4.11 shows how the accuracy of RP and PCA changes with increasing dimensions on the Caribou dataset. Starting from dimension two, RP immediately achieves high accuracy, while the accuracy of PCA reaches a local maxima at dimension four and then stays the same till the full dimension is attained. The reason we go upto 32 dimensions and then jump to the full dimension of 718 is because the PCA reports eigenvalues of magnitude zero beyond thirty two dimensions.

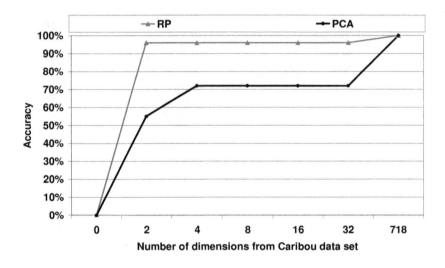

Figure 4.11: The accuracy of RP and PCA on Caribou dataset as a function of the dimension.

4.7   Summary and Conclusions

Given a large ST dataset, an important query is to retrieve objects which move close to each other for a long-duration of time. This is called a *Flock* pattern query and is the basis of many important ST pattern queries (Gudmundsson et al., 2007). The previously best known approximation algorithm for such queries have a running time which is either quadratic with respect to the number of entities or has an exponential dependency on the duration parameter of the query pattern. In this chapter we have proposed the use of random projection as a practical solution to manage the exponential dependency. We have proved that the random projection will return the "correct" answer with high probability, and our experiments on real, quasi-synthetic and synthetic datasets, strongly support our theoretical bounds. The use of random projection in conjunction with an indexing structure allows us to efficiently discover long flock patterns which was not possible till now.

Chapter 5

Mining Complex Time Series Patterns

This chapter proposes a novel technique to speed up the computation of DTW measure. It is organized as follows: We provide an introduction in Section 5.1. Section 5.2 describes related work on *DTW*. The *DTW* algorithm is described in Section 5.3. In Section 5.4, we give an overview of the techniques used to speed up DTW by adding constraints. Section 5.5 reviews the Divide and Conquer approach for *DTW* which is guaranteed to take up $O(m + n)$ space and $O(mn)$ time. Furthermore, we provide an example which clearly shows that the divide and conquer approach fails to arrive at the optimal *DTW* result. The *SparseDTW* algorithm is introduced with a detailed example in Section 5.6. In Section 5.7, we analyze and discuss our results. Table 5.1 lists the notations used in this chapter[1].

---

[1]This chapter is based on the following publications:

- Ghazi Al-Naymat, Sanjay Chawla and Javid Taheri. **SparseDTW: A Novel Approach to Speed up Dynamic Time Warping**. In communication with the Data and Knowledge Engineering (DKE), 2008 (Al-Naymat et al., 2008b).

Symbol	Description
TSD	Time Series Data
TSDM	Time Series Data Mining
DTW	Dynamic Time Warping
LCSS	Longest Common Subsequence
DC	Divide and Conquer
BandDTW	Sakoe-Chiba Band
SparseDTW	Sparse Dynamic Time Warping
EucDist	Euclidean Distance
$\mathcal{D}$	A set of time series (sequences) data
D	Warping Matrix
S	Time series (Sequence)
Q	Query time series (Sequence)
C	Candidate time series (Sequence)
s_i	The ith element of sequence S
q_i	The ith element of sequence Q
$DTW(Q, S)$	DTW distance between two time series Q and S
LBF	Lower Bound Function
$\lceil \cdot \rceil$	Function that gives the ceiling of a value
$\lfloor \cdot \rfloor$	Function that gives the floor of a value
$|S|$	The length of the time series S
SM	Sparse Matrix
D	Warping matrix
K	Number of cells in the sparse matrix
W	Warping path

Table 5.1: Description of the notations used.

5.1   Introduction

Dynamic time warping (*DTW*) uses the dynamic programming paradigm to compute the alignment between two time series. An *alignment* "warps" one time series onto another and can be used as a basis to determine the similarity between the time series. *DTW* has similarities to sequence alignment in bioinformatics and computational linguistics except that the *matching* process in sequence alignment and *warping* have to satisfy a different set of constraints and there is no gap condition in warping. *DTW* first became popular in the speech recognition community (Sakoe and Chiba, 1978) where it has been used to determine if the two speech wave-forms represent the same underlying spoken phrase. Since then it has been adopted in many other diverse areas and has become the similarity metric of choice in time series analysis (Keogh and Pazzani, 2000).

Like in sequence alignment, the standard *DTW* algorithm has $O(mn)$ space complexity where m and n are the lengths of the two sequences being aligned. This limits the practicality of the algorithm in todays "data rich environment" where long sequences are often the norm rather than the exception. For example, consider two time series which represent stock prices at one second granularity. A typical stock is traded for at least eight hours on the stock exchange and that corresponds to a length of $8 \times 60 \times 60 = 28800$. To compute the similarity, *DTW* would have to store a matrix with at least 800 million entries!

Figure 5.1(a) shows an example of an alignment (warping) between two sequences S and Q. It is clear that there are several possible alignments but the challenge is to select the one which has the minimal overall distance. The alignment has to satisfy several constraints which we will elaborate on in Section 5.3.

Salvador and Chan (2007) have provided a succinct categorization of different techniques that have been used to speed up *DTW*:

- **Constraints**: By adding additional constraints the search space of possible alignments can be reduced. Two well known exemplars of this approach are the Sakoe and Chiba (1978) and the Itakura (1975) constraints which limit how far the alignment can deviate from the diagonal. While these approaches provide a relief in the space complexity, they do not guarantee the optimality of the alignment.

- **Data Abstraction**: In this approach, the warping path is computed at a lower resolution of the data and then mapped back to the original resolution (Salvador and Chan, 2007). Again, optimality of the alignment is not guaranteed.

- **Indexing**: Keogh and Ratanamahatana (2004) and Sakurai et al. (2005) proposed an indexing approach, which does not directly speed up DTW but limits the number of DTW computations. For example, suppose there exists a database D of time series sequences and a query sequence q. We want to retrieve all sequences $d \in D$ such that $DTW(q, d) < \epsilon$. Then instead of checking q against each and every sequence in D, an easy to calculate lower bound function LBF is first applied between q and D. The argument works as follows:

  1. By construction, $LBF(q, d) < DTW(q, d)$.
  2. Therefore, if $LBF(q, d) > \epsilon$ then $DTW(q, d) > \epsilon$ and $DTW(q, d)$ does not have to be computed.

5.1.1 Main Contribution

The main insight behind our proposed approach, *SparseDTW*, is to dynamically exploit the possible existence of inherent similarity and correlation between the two time series whose DTW is being computed. This is the motivation behind the Sakoe-Chiba band and the Itakura Parellelogram but our approach has three distinct advantages:

1. Bands in *SparseDTW* evolve dynamically and are, on average, much smaller than the traditional approaches. We always represent the warping matrix using sparse matrices, which leads to better average space complexity compared to other approaches (Figure 5.9).

2. *SparseDTW* always yields the optimal warping path since we never have to set apriori constraints independently of the data. For example, in the traditional banded approaches, a sub-optimal path will result if all the possible optimal warping paths have to cross the bands.

3. Since *SparseDTW* yields an optimal alignment, it can easily be used in conjunction with lower bound approaches.

5.2   Related Work

DTW was first introduced in the data mining community in the context of mining time series (Berndt and Clifford, 1994). Since it is a flexible measure for time series similarity it is used extensively for ECGs (Electrocardiograms) (Caiani et al., 1998), speech processing (Rabiner and Juang, 1993), and robotics (Schmill et al., 1999). It is important to know that *DTW* is a measure not a metric, because *DTW* does not satisfy the triangular inequality.

Several techniques have been introduced to speed up *DTW* and/or reduce the space overhead (Hirschberg, 1975; Yi et al., 1998; Kim et al., 2001; Keogh and Ratanamahatana, 2004).

Divide and conquer (DC) heuristic proposed by Hirschberg (1975) is a dynamic programming algorithm that finds the least cost sequence alignment between two strings in linear space and quadratic time. The algorithm was first used in speech recognition area to solve the Longest Common Subsequence(LCSS). However as we will show with the help of an example, *DC* does not guarantee the optimality of the *DTW* distance.

Sakoe and Chiba (1978) speed up the *DTW* by constraining the warping path to lie within a band around the diagonal. However, if the optimal path crosses the band, the result will not be optimal.

Keogh and Ratanamahatana (2004) introduced an efficient lower bound that reduces the number of *DTW* computations in a time series database context. However itself it does not reduce the space complexity of the *DTW* computation, which is the objective our work.

Sakurai et al. (2005) presented FTW, a search method for DTW; it adds no global constraints on DTW. Their method designed based on a lower bounding distance measure that approximates the DTW distance. Therefore, it minimizes the number of DTW computations but does not increase the speed the DTW itself.

Salvador and Chan (2007) introduced an approximation algorithm for *DTW* called *Fast-DTW*. Their algorithm begins by using *DTW* in very low resolution, and progresses to a higher resolution linearly in space and time. *FastDTW* is performed in three steps: coarsening shrinks the time series into a smaller time series; the time series is projected by finding the minimum distance (warping path) in the lower resolution; and the warping path is an initial step for higher resolutions. The authors refined the warping path using local adjustment. *FastDTW* is an approximation algorithm, and thus there is no guarantee it will always find the optimal path. It requires the coarsening step to be run several times to produce many different resolutions of the time series. The *FastDTW* approach depends on a radius parameter as a constraint on the optimal path; however, our technique does not place any constrain while calculating the *DTW* distance.

DTW has been used in data streaming problems. Capitani and Ciaccia (2007) proposed a new technique, Stream-DTW (*STDW*). This measure is a lower bound of the *DTW*. Their method uses a sliding window of size 512. They incorporated a band constraint, forcing the path to stay within the band frontiers, as in (Sakoe and Chiba, 1978).

All the above algorithms were proposed either to speed up *DTW*, by reducing its space and time complexity, or reducing the number of *DTW* computations. Interestingly, the approach of exploiting the similarity between points (correlation) has never, to the best of our knowledge, been used in finding the optimality between two time series. *SparseDTW* considers the correlation between data points, that allows us to use a sparse matrix to

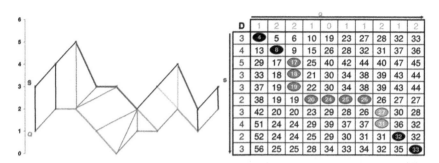

D	1	2	2	1	0	1	1	2	1	2
3	4	5	6	10	19	23	27	28	32	33
4	13	6	9	15	26	28	32	31	37	36
5	29	17	17	25	40	42	44	40	47	45
3	33	18	18	21	30	34	38	39	43	44
3	37	19	19	22	30	34	38	39	43	44
2	38	19	19	20	24	25	26	26	27	27
3	42	20	20	23	29	28	26	27	30	28
4	51	24	24	29	39	37	37	31	36	32
2	52	24	24	25	29	30	31	31	32	32
3	56	25	25	28	34	33	34	32	35	33

(a) The alignment of measurements for measuring the DTW distance between the two sequences S and Q.

(b) The warping matrix D produced by DTW; highlighted cells constitute the optimal warping path.

Figure 5.1: Illustration of *DTW*.

store the warping matrix instead of a full matrix. We do not believe that the idea of sparse matrix has been considered previously to reduce the required space.

Algorithm 5.1 *DTW*: The standard DTW algorithm.

Input: S: Sequence of length n, Q: Sequence of length m.
Output: *DTW distance*.
1: *Initialize $D(i, 1) \Leftarrow i\delta$ for each i*
2: *Initialize $D(1, j) \Leftarrow j\delta$ for each j*
3: **for all** i such that $2 \leq i \leq n$ **do**
4:     **for all** j such that $2 \leq j \leq m$ **do**
5:       *Use Equation 5.3 to compute $D(i, j)$*
6:     **end for**
7: **end for**
8: **return** $D(n, m)$

5.3   Dynamic Time Warping (DTW)

DTW is a dynamic programming technique used for measuring the similarity between any two time series with arbitrary lengths. This section gives an overview of *DTW* and how it is calculated. The following two time series will be used in our explanations.

$$S \; = \; s_1, s_2, s_3, \cdots, s_i, \cdots, s_n \tag{5.1}$$

$$Q \; = \; q_1, q_2, q_3, \cdots, q_j, \cdots, q_m \tag{5.2}$$

Where n and m represent the length of time series S and Q, respectively. i and j are the point indices in the time series.

DTW is a time series association algorithm that was originally used in speech recognition (Sakoe and Chiba, 1978). It relates two time series of feature vectors by warping the time axis of one series onto another.

As a dynamic programming technique, it divides the problem into several sub-problems, each of which contribute in calculating the distance cumulatively. Equation 5.3 shows the recursion that governs the computations is:

$$D(i,j) = d(i,j) + min \begin{cases} D(i-1,j) \\ D(i-1,j-1) \\ D(i,j-1). \end{cases} \tag{5.3}$$

The first stage in the *DTW* algorithm is to fill a local distance matrix d. That matrix has $n \times m$ elements which represent the Euclidean distance between every two points in the time series (i.e., distance matrix). In the second stage, it fills the warping matrix D (Figure 5.1(b)) on the basis of Equation 5.3. Lines 1 to 7 in Algorithm 5.1 illustrate the process of filling the warping matrix. We refer to the cost between the ith and the jth

elements as δ as mentioned in line 1 and 2.

After filling the warping matrix, the final stage for the *DTW* is to report the optimal warping path and the *DTW* distance. Warping path is a set of adjacent matrix elements that identify the mapping between S and Q. It represents the path that minimizes the overall distance between S and Q. The total number of elements in the warping path is K, where K denotes the normalizing factor and it has the following attributes:

$$W = w_1, w_2, \dots, w_K$$

$$max(|S|, |Q|) \leq K < (|S| + |Q|)$$

Every warping path must satisfy the following constraints (Keogh and Ratanamahatana, 2004; Salvador and Chan, 2007; Sakoe and Chiba, 1978):

1. **Monotonicity:** Any two adjacent elements of the warping path W, $w_k = (w_i, w_j)$ and $w_{k-1} = (w_i', w_j')$, follow the inequalities, $w_i - w_i' \geq 0$ and $w_j - w_j' \geq 0$. This constrain guarantees that the warping path will not roll back on itself. That is, both indexes i and j either stay the same or increase (they never decrease).

2. **Continuity:** Any two adjacent elements of the warping path W, $w_k = (w_i, w_j)$ and $w_{k+1} = (w_i', w_j')$, follow the inequalities, $w_i - w_i' \leq 1$ and $w_j - w_j' \leq 1$. This constraint guarantees that the warping path advances one step at a time. That is, both indexes i and j can only increase by at most 1 on each step along the path.

3. **Boundary:** The warping path starts from the top left corner $w_1 = (1, 1)$ and ends at the bottom right corner $w_k = (n, m)$. This constraint guarantees that the warping path contains all points of both time series.

Although there are a large number of warping paths that satisfy all of the above constraints, *DTW* is designed to find the one that minimizes the warping cost (distance).
Figures 5.1(a) and 5.1(b) demonstrate an example of how two time series (S and Q) are
warped and the way their distance is calculated. The circled cells show the optimal warping path, which crosses the grid from the top left corner to the bottom right corner. The
DTW distance between the two time series is calculated based on this optimal warping
path using the following equation:

$$DTW(S,Q) = min \left\{ \frac{\sqrt{\sum_{k=1}^{K} W_k}}{K} \right. \tag{5.4}$$

The K in the denominator is used to normalize different warping paths with different
lengths.

Since the *DTW* has to potentially examine every cell in the warping matrix, its space and
time complexity is $O(nm)$.

5.4   Global Constraint (BandDTW)

In Chapter 2 we described several methods that add global constraints on DTW to increase its speed by limiting how far the warping path may stray from the diagonal of
the warping matrix. In this chapter we use Sakoe-Chiba Band (henceforth, we refer to
it as BandDTW) Sakoe and Chiba (1978) when comparing with our proposed algorithm
(Figure 5.2). BandDTW used to speed up the *DTW* by adding constraints which force
the warping path to lie within a band around the diagonal; if the optimal path crosses the
band, the DTW distance will not be optimal.

Figure 5.2: Global constraint (Sakoe Chiba Band), which limits the warping scope. The diagonal green areas correspond to the warping scopes.

5.5 Divide and Conquer Technique (DC)

In Section 5.3, we have shown how to compute the optimal alignment using the standard *DTW* technique between two time series. In this section we will show another technique that uses a Divide and Conquer heuristic, henceforth we refer to it as (*DC*), proposed by Hirschberg (1975). *DC* is a dynamic programming algorithm used to find the least cost sequence alignment between two strings. The algorithm was first introduced to solve the Longest Common Subsequence (LCSS) (Hirschberg, 1975). Algorithm 5.2 gives a high level description of the *DC* algorithm. Like in the standard sequence alignment, the *DC* algorithm has $O(mn)$ time complexity but $O(m+n)$ space complexity, where m and n are the lengths of the two sequences being aligned. We will be using Algorithm 5.2 along with Figure 5.3 to explain how *DC* works. In the example we use two sequences $S = [3, 4, 5, 3, 3]$ and $Q = [1, 2, 2, 1, 0]$ to determine the optimal alignment between

---

Algorithm 5.2 *DC*: Divide and Conquer technique.

Input: *S: Sequence of length n, Q: Sequence of length m.*
Output: *DTW distance.*
 1: *Divide-Conquer-Alignment(S,Q)*
 2: $n \Leftarrow |S|$
 3: $m \Leftarrow |Q|$
 4: $Mid \Leftarrow \lceil m/2 \rceil$
 5: **if** $n \leq 2$ or $m \leq 2$ **then**
 6:     *Compute optimal alignment using standard DTW*
 7: **else**
 8:     $f \Leftarrow ForwardsSpaceEfficientAlign(S,Q[1:Mid])$
 9:     $g \Leftarrow BackwardsSpaceEfficientAlign(S,Q[Mid:m])$
10:     $q \Leftarrow$ *index that minimizing* $f(q, Mid) + g(q, Mid)$
11:     *Add (q,Mid) to global array P*
12:     *Divide-Conquer-Alignment(S[1:q],Q[1:Mid])*
13:     *Divide-Conquer-Alignment(S[q:n],Q[Mid:m])*
14: **end if**
15: **return** *P*

---

them. There is only one optimal alignment for this example (Figure 5.3(e)), where shaded cells are the optimal warping path. The *DC* algorithm works as follows:

1. It finds the middle point in Q which is $Mid = |Q|/2$, (Figure 5.3(a)). This helps to find the split point which divides the warping matrix into two parts (sub-problems). A forward space efficiency function (Line 8) uses S and the first cut of $Q = [1, 2, 2]$, then a backward step (Line 9) uses S and $Q = [2, 1, 0]$ (Figure 5.3(a)). Then by adding the last column from the forward and backward steps together and finding the index of the minimum value, the resultant column indicates the row index that will be used along with the middle point to locate the split point (shaded cell in Figure 5.3(a)). Thus, the first split point is D(4,3). At this stage of the algorithm, there are two sub-problems; the alignment of $S = [3, 4, 5, 3]$ with $Q = [1, 2, 2]$ and of $S = [3, 3]$ with $Q = [2, 1, 0]$.

2. *DC* is recursive algorithm, each call splits the problem into two other sub-problems if both sequences are of $length > 2$, otherwise it calls the standard *DTW* to find

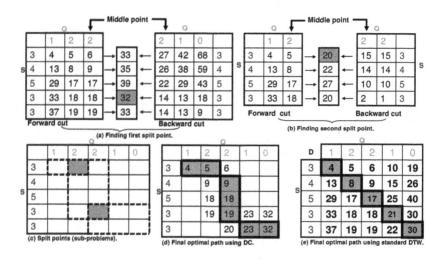

Figure 5.3: An example to show the difference between the standard DTW and the DC algorithm.

the optimal path for that particular sub-problem. In the example, the first sub-problem will be fed to Line 12 which will find another split point, because both input sequences are of length > 2. Figure 5.3(b) shows how the new split point is found. Figure 5.3(c) shows the two split points (shaded cells) which yield to have sub-problems of sequences of length ≤ 2. In this case DTW will be used to find the optimal alignment for each sub-problem.

3. The DC algorithm finds the final alignment by concatenating the results from each call of the standard DTW.

The example in Figure 5.3 clarifies that the DC algorithm does not give the optimal warping path. Figures 5.3(d) and (e) show the paths obtained by the DC and DTW algorithms, respectively.

DC does not yield the optimal path as it goes into infinite recursion because of how it calculates the middle point. DC calculates the middle point as follows:

There are two scenarios: first, when the middle point (Algorithm 5.2 Line 4) is floored ($Mid = \lfloor m/2 \rfloor$) and second when it is rounded up ($Mid = \lceil m/2 \rceil$). The first scenario causes infinite recursion, since the split from the previous step gives the same subsequences (i.e., the algorithm keeps finding the same split point). The second scenario is shown in Figures 5.3(a-d), which clearly confirms that the final optimal path is not the same as the one retrieved by the standard DTW [2]. The final DTW distance is different as well. The shaded cells in Figures 5.3(d) and (e) show that both warping paths are different.

5.6   Sparse Dynamic Programming Approach

In this section we outline the main principles we use in *SparseDTW* and follow up with an illustrated example along with the *SparseDTW* pseudo-code. We exploit the following facts in order to reduce space usage while avoiding any re-computations:

1. Quantizing the input time series to exploit the similarity between the points in the two time series.

2. Using a sparse matrix of size k, where $k = n \times m$ in the worst case. However, if the two sequences are similar, $k << n \times m$.

3. The warping matrix is calculated using dynamic programming and sparse matrix indexing.

---

[2]It should be noted that our example has only one optimal path that gives the optimal distance.

5.6.1 Key Concepts

In this section we introduce the key concepts used in our algorithm.

Definition 5.1 (Sparse Matrix SM) *is a matrix that is populated largely with zeros. It allows the techniques to take advantage of the large number of zero elements. Figure 5.4(a) shows the SM initial state. SM is linearly indexed, The little numbers, in the top left corner of SM's cells, represent the cell index. For example, the indices of the cells $SM(1,1)$ and $SM(5,5)$ are 1 and 5, respectively.*

Definition 5.2 (Lower Neighbors ($LowerNeighbors$)) *a cell $c \in SM$ has three lower neighbors which are the cells of the indices $(c-1)$, $(c-n)$, and $(c-(n+1))$ (where n is the number of rows in SM). For example, the index of cell $SM(2,3)$ is 12 and its lower neighbors are $SM(1,3)$, $SM(1,2)$ and $SM(2,2)$ which has indices 11,6 and 7, respectively (Figure 5.4(a)).*

Definition 5.3 (Upper Neighbors ($UpperNeighbors$)) *a cell $c \in SM$ has three upper neighbors which are the cells of the indices $(c+1)$, $(c+n)$, and $(c+n+1)$ (where n is the number of rows in SM). For example, the index of cell $SM(2,3)$ is 12 and its upper neighbors are $SM(3,3)$, $SM(2,4)$ and $SM(3,4)$ which has indices 13,17 and 18, respectively (Figure 5.4(a)).*

Definition 5.4 (Blocked Cell (B)) *a cell $c \in SM$ is blocked if its value is zero. The letter (B) refers to the blocked cells (Figure 5.4(a)).*

Definition 5.5 (Unblocking) *Given a cell $c \in SM$, if $SM(c)$'s upper neighbors ($SM(c+1), SM(c+n)$, and $SM(c+n+1)$) are blocked, they will be unblocked. Unblocking is performed by calculating the EucDist for these cells and adding them to SM. In*

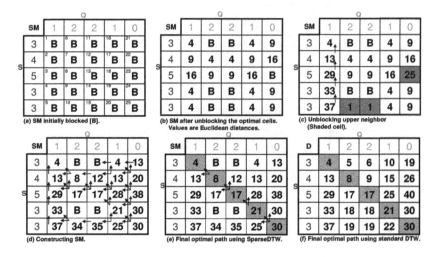

Figure 5.4: An example of the SparseDTW algorithm and the method of finding the optimal path.

other words, adding the distances to these cells means changing their state from blocked (B) into unblocked. For example, the upper neighbors of $SM(4,2)$ (Figure 5.4(c)) are blocked. They need to be unblocked.

5.6.2 SparseDTW Algorithm

The algorithm takes Res, the resolution parameter as an input that determines the number of bins as $\frac{2}{Res}$. Res will have no impact on the optimality. We now present an example of our algorithm to illustrate some of the highlights of our approach: We start with two sequences:

$$S = [3, 4, 5, 3, 3] \text{ and } Q = [1, 2, 2, 1, 0].$$

In Line 1, we first quantize the sequences into the range $[0, 1]$ using Equation 5.5:

Algorithm 5.3 *SparseDTW*: Sparse dynamic programming technique.

Input: *S: Time series of length* n*, Q: Time series of length* m*, and Res.*
Output: *Optimal warping path* and *SparseDTW* distance.

1: $[S', Q'] \Leftarrow Quantize(S, Q)$
2: $LowerBound \Leftarrow 0, UpperBound \Leftarrow Res$
3: **for all** $0 \leq LowerBound \leq 1 - \frac{Res}{2}$ **do**
4:       $IdxS \Leftarrow find(LowerBound \leq S' \leq UpperBound)$
5:       $IdxQ \Leftarrow find(LowerBound \leq Q' \leq UpperBound)$
6:       $LowerBound \Leftarrow LowerBound + \frac{Res}{2}$
7:       $UpperBound \Leftarrow LowerBound + Res$
8:       **for all** $idx_i \in IdxS$ **do**
9:           **for all** $idx_j \in IdxQ$ **do**
10:              Add $EucDist(idx_i, idx_j)$ to SM {When $EucDist(idx_i, idx_j) = 0, SM(i, j) = -1.$}
11:          **end for**
12:      **end for**
13: **end for**
     {Note: SM is linearly indexed.}
14: **for all** $c \in SM$ **do**
15:      $LowerNeighbors \Leftarrow \{(c - 1), (c - n), (c - (n + 1))\}$
16:      $minCost \Leftarrow min(SM(LowerNeighbors))$ {SM(LowerNeighbors)=-1 means cost=0.}
17:      $SM(c) \Leftarrow SM(c) + minCost$
18:      $UpperNeighbors \Leftarrow \{(c + 1), (c + n), (c + n + 1)\}$
19:      **if** $|UpperNeighbors| == 0$ **then**
20:          $SM \cup EucDist(UpperNeighbors)$
21:      **end if**
22: **end for**
23: $WarpingPath \Leftarrow \Phi$
24: $hop \Leftarrow SM(n \times m)$ {Last index in **SM**.}
25: $WarpingPath \cup hop$
26: **while** $hop \neq SM(1)$ **do**
27:      $LowerNeighbors \Leftarrow \{(hop - 1), (hop - n), (hop - (n + 1))\}$
28:      $[minCost, index] \Leftarrow min[Cost([LowerNeighbors])]$
29:      $hop \Leftarrow index$
30:      $WarpingPath \cup hop$
31: **end while**
32: $WarpingPath \cup SM(1)$
33: **return** $WarpingPath, SM(n \times m)$

$$QuantizedSeq_i^k = \frac{S_i^k - min(S^k)}{max(S^k) - min(S^k)}. \tag{5.5}$$

Where S_i^k denotes the ith element of the kth time series. This yields the following sequences:

$S' = [0, 0.5, 1.0, 0.0, 0.0]$ and $Q' = [0.5, 1.0, 1.0, 0.5, 0]$

In Lines 4 to 7 we create *overlapping* bins, governed by two parameters: bin-width and the overlapping width (which we refer to as the resolution). It is important to note that these two parameters do not affect the optimality of the alignment but do have an affect on the amount of space utilized. For this particular example, the bin-width is 0.5. We thus have 4 bins which are shown in Table 5.2. We iterate over these bins by changing the variable LowerBound.

Bin Number (B_k)	Bin Bounds	Indices Bin of S'	Indices of Q'
1	0.0-0.5	1,2,4,5	1,4,5
2	0.25-0.75	2	1,4
3	0.5-1.0	2,3	1,2,3,4
4	0.75-1.25	3	2,3

Table 5.2: Bins bounds, where B_k is the kth bin.

Our intuition is that points in sequences with similar profiles will be mapped to other points in the same bin or neighboring bins. In which case the non-default entries of the sparse matrix can be used to compute the warping path. Otherwise, default entries of the matrix will have to be "opened", reducing the sparsity of the matrix but never sacrificing the optimal alignment.

In Lines 3 to 13, the sparse warping matrix SM is constructed using the equation below. SM^3 is a matrix that has generally few non-zero (or "interesting") entries. It can be

---

[3]If the Euclidean distance (EucDist) between $S(i)$ and $Q(j)$ is zero, then $SM(i, j) = -1$, to distinguish

represented in much less than $n \times m$ space, where n and m are the lengths of the time series S and Q, respectively.

$$SM(i,j) = \begin{cases} EucDist(S(i), Q(j)) & \text{if } S(i) \text{ and } Q(j) \in B_k \\ B & otherwise \end{cases} \tag{5.6}$$

We assume that SM is linearly ordered and the default value of SM cells are zeros. That means the cells initially are *Blocked* (B) (Figure 5.4(a)). Figure 5.4(a) shows the linear order of the SM matrix, where the little numbers on the top left corner of each cell represent the index of the cells. In Line 6 and 7, we find the index of each quantized value that falls in the bin bounds (Table 5.2 column 2, 3 and 4). The Inequality 5.7 is used in Line 4 and 5 to find the indices of the default entries of the SM.

$$LowerBound \leq QuantizedSeq_i^k \leq UpperBound. \tag{5.7}$$

Where $LowerBound$ and $UpperBound$ are the bin bounds and $QuantizedSeq_i^k$ represents the quantized time series which can be calculated using Equation 5.5.

Lines 8 to 12 are used to initialize the SM. That is by joining all indices in $idxS$ and $idxQ$ to open corresponding cells in SM. After unblocking (opening) the cells that reflect the similarity between points in both sequences, the SM entries are shown in Figure 5.4(b).

Lines 14 to 22 are used to calculate the warping cost. In Line 15, we find the warping cost for each open cell $c \in SM$ (cell c is the number from the linear order of SM's cells) by finding the minimum of the costs of its lower neighbors, which are $[c-1, c-n, c-(n+1)]$ (black arrows in Figure 5.4(d) show the lower neighbors of every open cell). This cost is then added to the local distance of cell c (Line 17). The above step is similar to *DTW*,

---

between a blocked cell and any cell that represents zero distance.

however, we may have to open new cells if the upper neighbors at a given local cell $c \in SM$ are blocked. The indices of the upper neighbors are $[c + 1, c + n, c + n + 1]$, where n is the length of sequence S (i.e., number of rows in SM). Lines 18 to 21 are used to check always the upper neighbors of $c \in SM$. This is performed as follows: if the $|UpperNeighbors| = 0$ for a particular cell, its upper neighbors will be unblocked. This is very useful when the algorithm traverses SM in reverse to find the final optimal path. In other words, unblocking allows the path to be connected. For example, the cell $SM(5, 1)$ has one upper neighbor that is cell $SM(5, 2)$ which is blocked (Figure 5.4(c)), therefore this cell will be unblocked by calculating the EucDist(S(5),Q(2)). The value will be add to the SM which means that cell $SM(5, 2)$ is now an entry in SM (Figure 5.4(c)). Although unblocking adds cells to SM which means the number of open cells will increase, but the overlapping in the bins boundaries allows the SM's unblocked cells to be connected mostly that means less number of unblocking operations. Figure 5.4(d) shows the final entries of the SM after calculating the warping cost of all open cells.

Lines 23 to 32 return the warping path. hop initially represents the linear index for the (m, n) entry of SM, that is the bottom right corner of SM in Figure 5.4(e). Starting from $hop = n \times m$ we choose the neighbors $[hop - n, hop - 1, hop - (n + 1)]$ with minimum warping cost and proceed recursively until we reach the first entry of SM, namely, $SM(1, 1)$ or $hop = 1$. It is interesting that while calculating the warping path we only have to look at the open cells, which may be fewer in number than 3. This potentially reduces the overall time complexity.

Figure 5.4(e) demonstrates an example of how the two time series (S and Q) are warped and the way their distance is calculated using *SparseDTW*. The filled cells show the optimal warping path, which crosses the grid from the top left corner to the bottom right corner. The distance between the two time series is calculated using Equation 5.4. Figure 5.4(f) shows the standard *DTW* where the filled cells are the optimal warping path. It is clear that both techniques give the optimal warping path which will yield the optimal distance.

5.6.3 SparseDTW Complexity

Given two time series S and Q of length n and m, the space and time complexity of standard *DTW* is $O(nm)$. For *SparseDTW* we attain a reduction by a constant factor b, where b is the number of bins. This is similar to the *BandDTW* approach where the reduction in space complexity is governed by the size of the band. However, *SparseDTW* always yields the optimal alignment. The time complexity of *SparseDTW* is $O(nm)$ in the worst case as we potentially have to access every cell in the matrix.

5.7 Experiments, Results and Analysis

In this section we report and analyze the experiments that we have conducted to compare *SparseDTW* with other methods. Our main objective is to evaluate the space-time trade-off between *SparseDTW*, *BandDTW* and *DTW*. We evaluate the effect of *correlation* on the running time of *SparseDTW*[4]. As we have noted before, both *SparseDTW* and *DTW* always yield the optimal alignment while *BandDTW* results can often lead to sub-optimal alignments, as the optimal warping path may lie outside the band. As we noted before *DC* may not yield the optimal result.

---

[4] The run time includes the time used for constructing the Sparse Matrix SM.

5.7.1   Experimental Setup

All experiments were carried out on a PC with Windows XP operating system, a Pentium(R) D (3.4 GHz) processor and 2 GB main memory. The data structures and algorithm were implemented in C++.

5.7.2   Datasets

We have used a combination of benchmark and synthetically generated datasets. The benchmark dataset is a subset from the *UCR* time series data mining archive (Keogh, 2006). We have also generated synthetic time series data to control and test the effect of correlation on the running time of *SparseDTW*. Appendix C provides more details about the preparation process and the source of the datasets. We briefly describe the characteristics of each dataset used.

- **GunX:** comes from the video surveillance application and captures the shape of a gun draw with the gun in hand or just using the finger. The shape is captured using 150 time steps and there are a total of 100 sequences (Keogh, 2006). We randomly selected two sequences and computed their similarity using the three methods.

- **Trace:** is a synthetic dataset generated to simulate instrumentation failures in a nuclear power plant (Roverso, 2000). The dataset consists of 200 time series each of length 273.

- **Burst-Water:** is formed by combining two different datasets from two different applications. The average length of the series is 2200 points (Keogh, 2006).

- **Sun-Spot:** is a large dataset that has been collected since 1818. We have used the daily sunspot numbers. More details about this dataset exists in (Vanderlinden,

2008). The 1st column of the data is the year, month and day, the 2nd column is year and fraction of year (in Julian year)[5], and the 3rd column is the sunspot number. The length of the time series is 2898.

- **ERP:** is the Event Related Potentials that are calculated on human subjects[6]. The dataset consists of twenty sequences of length 256 (Makeig et al., 1999).

- **Synthetic:** Synthetic datasets were generated to control the correlation between sequences. The length of each sequence is 500.

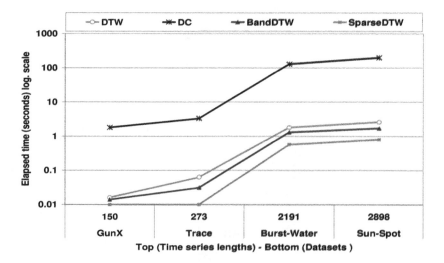

Figure 5.5: Elapsed time using real life datasets.

---

[5]The Julian year is a time interval of exactly 365.25 days, used in astronomy.
[6]An indirect way of calculating the brain response time to certain stimuli.

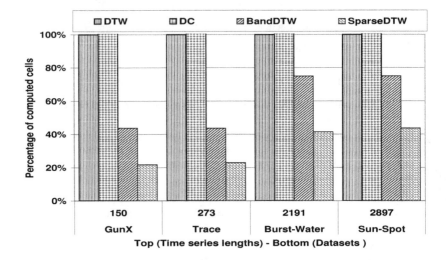

Figure 5.6: Percentage of computed cells as a measure for time complexity.

| Data size | Number of computed cells used by | | | |
	DTW	DC	BandDTW	SparseDTW
2K	4×10^6	$> 8 \times 10^6$	2500	2000
4K	16×10^6	$> 30 \times 10^6$	5000	4000
6K	36×10^6	$> 70 \times 10^6$	7500	6000

Table 5.3: Number of computed cells if the optimal path is close to the *diagonal*.

5.7.3   Discussion and Analysis

SparseDTW algorithm is evaluated against three other existing algorithms, *DTW*, which always gives the optimal answer, *DC*, and *BandDTW*.

5.7.3.1   Elapsed Time

The running time of the four approaches is shown in Figure 5.5. The time profile of both *DTW* and *BandDTW* is similar and highlights the fact that *BandDTW* does not exploit

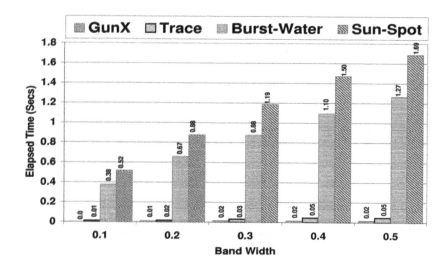

Figure 5.7: Effect of the band width on *BandDTW* elapsed time.

Dataset size	Algorithm name	#opened cells	Elapsed Time(Sec.)
3K	DTW	9×10^6	7.3
	SparseDTW	614654	0.65
6K	DTW	36×10^6	26
	SparseDTW	2048323	2.2
9K	DTW	81×10^6	N.A
	SparseDTW	4343504	4.8
12K	DTW	144×10^6	N.A
	SparseDTW	7455538	200

Table 5.4: Performance of the *DTW* and *SparseDTW* algorithms using large datasets.

the nature of the datasets. *DC* shows as well the worst performance due to the vast number of recursive calls to generate and solve sub-problems. In contrast, it appears that *SparseDTW* is exploiting the inherent similarity in the GunX and Trace data.

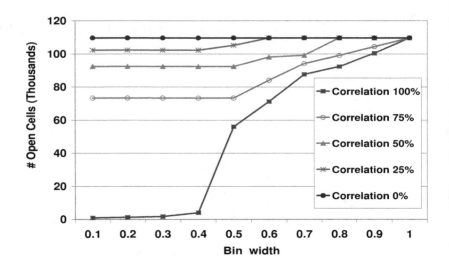

Figure 5.8: Effects of the resolution and correlation on *SparseDTW*.

In Figure 5.6 we show the number of open/computed cells produced by the four algorithms. It is very clear that *SparseDTW* produces the lowest number of opened cells.

In Table 5.3 we show the number of computed cells that are used in finding the optimal alignment for three different datasets, where their optimal paths are close to the diagonal. *DC* has shown the highest number of computed cells followed by *DTW*. That is because both (*DC* and *DTW*) do not exploit the similarity in the data. *BandDTW* has shown interesting results here because the optimal alignment is close to the diagonal. However, *SparseDTW* still outperforms it.

Two conclusions are revealed from Figure 5.7. The first, the length of the time series affects the computing time, because the longer the time series the bigger the matrix. Second, band width influences CPU time when aligning pairs of time series. The wider the band the more cells are required to be opened.

Dataset name	Algorithm name	Number of opened cells	Warping path size (K)	Elapsed Time (Seconds)	DTW Distance
GunX	DTW	22500	201	0.016	0.01
	BandDTW	448	152	0.000	0.037
	SparseDTW	4804	201	0.000	0.01
Trace	DTW	75076	404	0.063	0.002
	BandDTW	1364	331	0.016	0.012
	SparseDTW	17220	404	0.000	0.002
Burst-Water	DTW	2190000	2190	1.578	0.102
	BandDTW	43576	2190	0.11	0.107
	SparseDTW	951150	2190	0.75	0.102
Sun-Spot	DTW	1266610	357	0.063	0.021
	BandDTW	12457	358	0.016	0.022
	SparseDTW	66049	357	0.016	0.021
ERP	DTW	1000000	1533	0.78	0.008
	BandDTW	19286	1397	0.047	0.013
	SparseDTW	210633	1535	0.18	0.008
Synthetic	DTW	250000	775	0.187	0.033
	BandDTW	4670	600	0.016	0.043
	SparseDTW	105701	775	0.094	0.033

Table 5.5: Statistics about the performance of *DTW*, *BandDTW*, and *SparseDTW*. Results in this table represent the average over all queries.

DTW and *SparseDTW* are compared together using large datasets. Table 5.4 shows that *DTW* is not applicable (N.A) for datasets of size $> 6K$, since it exceeds the size of the memory when computing the warping matrix. In this experiment we excluded *BandDTW* and *DC* given that they provide no guarantee on the optimality.

To determine the effect of correlation on the elapsed time for *SparseDTW* we created several synthetic datasets with different correlations. The intuition being that two sequences with lower correlation will have a warping path which is further away from the diagonal and thus will require more open cells in the warping matrix. The results in Figure 5.8 confirm our intuition though only in the sense that extremely low correlation sequences have a higher number of open cells than extremely high correlation sequences.

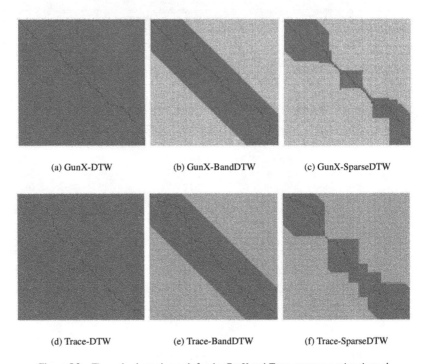

(a) GunX-DTW (b) GunX-BandDTW (c) GunX-SparseDTW

(d) Trace-DTW (e) Trace-BandDTW (f) Trace-SparseDTW

Figure 5.9: The optimal warping path for the GunX and Trace sequences using three al-gorithms (*DTW*, *BandDTW*, and *SparseDTW*). The advantages of *SparseDTW* are clearly revealed as only a small fraction of the matrix cells have to be "opened" compared to the other two approaches.

5.7.3.2 SparseDTW Accuracy

The accuracy of the warping path distance of *BandDTW* and *SparseDTW* compared to standard *DTW* (which always gives the optimal result) is shown in Table 5.5. It is clear that the error rate of *BandDTW* varies from 30% to 500% while *SparseDTW* always gives the exact value. It should be noticed that there may be more than one optimal path of different sizes but they should give the same minimum cost (distance). For example, the size of the warping path for the *ERP* dataset produced by *DTW* is 1533, however, *SparseDTW* finds another path of size 1535 with the same distance as *DTW*.

Figure 5.9 shows the dramatic nature in which *SparseDTW* exploits the similarity inherent in the sequences and creates an adaptive band around the warping path. For both the GunX and the Trace data, *SparseDTW* only opens a fraction of the cells compared to both standard *DTW* and *BandDTW*.

5.8 Summary and Conclusions

In this chapter we have introduced the *SparseDTW* algorithm, which is a sparse dynamic programming technique. It exploits the correlation between any two time series to find the optimal warping path between them. The algorithm finds the optimal path efficiently and accurately. *SparseDTW* always outperforms the algorithms *DTW*, *BandDTW* and *DC*. We have shown the efficiency of the proposed algorithm through comprehensive experiments using synthetic and real life datasets. Our algorithm can be easily combined with lower bound techniques. The SparseDTW algorithm was applied on large stock market data to efficiently find pairs trading patterns (Chapter 6).

Chapter 6

Pairs Trading Mining using SparseDTW

We propose an efficient approach for reporting pairs trading patterns in large stock market datasets. This will be achieved by using our proposed algorithm (SparseDTW), presented in Chapter 5. In Section 6.1, we provide an overview about pairs trading strategy, and define the key concepts. A review of state-of-the-art research conducted in mining pairs trading is given in Section 6.2. Section 6.3 demonstrates the pairs trading framework. Several classical methods that are used to find the pairs trading patterns are described in Section 6.4. This section also describes how the speed of a SparseDTW sequential search can be increased using a lower bounding technique. Table 6.1 lists the notations used in this chapter[1].

---

[1]This chapter is based on the following publications:

- Ghazi Al-Naymat, Sanjay Chawla and Javid Taheri. **SparseDTW: A Novel Approach to Speed up Dynamic Time Warping**. In communication with the journal of Data and Knowledge Engineering (DKE), 2008 (Al-Naymat et al., 2008b).

Symbol	Description
TSD	Time Series Data
TSDM	Time Series Data Mining
DTW	Dynamic Time Warping
EucDist	Euclidean Distance
SparseDTW	Sparse Dynamic Time Warping
S	Time series (Sequence)
Q	Query time series (Sequence)
C	Candidate time series (Sequence)
s_i	The ith element of sequence S
q_i	The ith element of sequence Q
P_i	The price of stock i
nP_i	Normalized price of stock i
$DTW(Q, S)$	DTW distance between two time series Q and S
$LB\_Keogh(Q, C)$	Lower bound for query sequence Q and candidate sequence C
SP	Stock pair
ASX	Australian Stock eXchange

Table 6.1: Description of the notations used.

6.1   Introduction

The volume of financial data has rapidly increased due to advances in software and hardware technologies. Stock markets provide examples of financial data that contains many attributes – far more than traders can readily understand. Traders nonetheless attempt to determine relationships between data attributes that can yield profitable trading of financial instruments. As traders' needs have become more complex, the demand for more efficient techniques has grown. Many researchers have developed algorithms and frameworks that concentrate on mining useful patterns in stock market datasets. Interesting patterns include *collusion, money laundering, insider trading*, and *pairs trading* (Donoho, 2004; Zhang et al., 2003; Little et al., 2002; Gatev et al., 2006; Vidyamurthy, 2004; Nesbitt and Barrass, 2004). The literature has shown that pairs trading is one of the most sought-after patterns because of its market-neutral strategy – the return is uncorrelated to the market (Vidyamurthy, 2004). The concept of pairs trading was originally developed in the late 1980s by quantitative analysts and was subsequently refined by Gerald Bamberger at Morgan Stanley during the 1990s[2]. With the help of others, they found that the day-to-day price movements of certain securities, often competitors in the same sector, are closely correlated. When the correlation breaks down, i.e., one stock trades up while the other trades down, traders tend to sell the outperforming stock and buy the underperforming one, anticipating that the "spread"[3] between the two will eventually converge. A number of real-life examples of potentially correlated pairs are listed below (Vidyamurthy, 2004; Nesbitt and Barrass, 2004).

- Coca-Cola (KO) and Pepsi (PEP)

- Wal-Mart (WMT) and Target Corporation (TGT)

- Dell (DELL) and Hewlett-Packard (HPQ)

- Ford (F) and General Motors (GM)

Pairs trading is an investment strategy that involves buying the undervalued security and short-selling the overvalued one, thus maintaining market neutrality. This helps to hedge sector and market risk. For example, if the market crashes and a trader's two stocks plummet with it, the gain will occur on the short position and lose on the long position, which minimizes the overall loss. Finding pairs trading is one of the pivotal issues in the stock market, because investors tend to conceal from others their prior knowledge about the stocks that form pairs, to gain the greatest advantage from them; in other words, investors always try to selfishly exploit market inefficiency. The idea behind pairs trading is to profit from market amendments towards the normal behavior. To this end, the reason to understand and identify pairs trading is to help all investors take advantage of the large number of stocks that appear in pairs. This can guide them to invest their money in stocks

---

[2](NYSE: MS) is a global financial services provider headquartered in New York City.
[3]Spread is the difference between bid and ask prices.

that have a lower market risk and to return the maximum profit (i.e., by guiding investors to choose the right time to buy and sell particular stocks) (Vidyamurthy, 2004; Gatev et al., 2006; Cao et al., 2006b).

To find the similarity between time series, a similarity measure should be used. Examples of these measures include Euclidean distance and Dynamic Time Warping (DTW). Previous research in this area concentrated in using the Euclidean distance as the similarity measure when comparing stock prices. As shown in Chapter 2, the Euclidean distance is not a good measure for similarity due to its limitations – it does not, for example, capture the shape of the time series because of its linearity method of calculating the distance. In this research, we used the DTW as the underlying similarity measure. DTW is a well-accepted measure, due to its ability to capture the time series shapes as well as its power to handle time series of different lengths. The space and time complexity of DTW is $O(nm)$, where n and m are the length of the time series being compared – this is the major limitation DTW. Therefore, we have used our novel technique (SparseDTW), which increases the speed of DTW by reducing its space complexity using sparse matrices, as described in Chapter 5. This is the first time that DTW has been used for mining pairs trading patterns from large stock-market data. Our experiments show the validity of the proposed framework.

6.1.1 Problem Statement

Given a large time series dataset (stock market data set), efficiently find all stock pairs that are correlated (similar) as defined by their similarity measure (DTW), being below a threshold.

6.1.2 Contributions

In this chapter we make the following contributions:

1. We propose a framework to find efficiently pairs trading patterns amongst large stock market datasets.

2. We show for the first time the use of DTW as a similarity measure in finding pairs trading patterns.

3. We show that our proposed technique (SparseDTW) can be used to mine stock pairs successfully, and can be combined with a lower bounding technique to reduce the number of its computation.

4. We describe two different trading rules which can be used by analysts to guide investors to exploit periods of opportunity for generating profit, and to reduce the market risk.

6.1.3 Key Concepts

Definition 6.1 (Short selling) *is the practice of selling stocks which are not owned by the seller, in the hope of repurchasing them in the future at a lower price.*

Definition 6.2 (Long buying) *is the practice of buying stocks in the hope of selling them in the future at a higher price.*

Definition 6.3 (Ask price) *is the lowest price a seller of a stock is willing to accept for a share of a given stock. .*

Definition 6.4 (Bid price) *is the highest price that a buyer is willing to pay for a share of a given stock.*

Definition 6.5 (Spread) *is the difference between the price available for an immediate sale (a "bid") and an immediate purchase (an "ask").*

Definition 6.6 (Pairs trading) *is a strategy to find two stocks whose prices have moved together over a period of time. Once the price deviation is noticed, two opposite positions will be taken to make the most profit (by short-selling one and long-buying the other) (Gatev et al., 2006; Vidyamurthy, 2004).*

6.2   Related Work

Mining pairs has attracted the attention of the data mining and machine learning community for the last decade. A number of algorithms have been proposed for extracting knowledge from stock market data sets. This section reviews the most recent techniques that have been developed to mine pairs trading.

Stock market prediction has been a major issue in the field of finance. Neural networks (NNs) were used to mitigate the prediction issue. The most primitive stock market prediction model based on NNs was designed by White (White, 1988; WEIGEND, 1996). They used feed forward neural networks (FFNNs) to interpret previously hidden regularities in the equity price movements, such as oscillations of stock prices, and showed how to search for such regularities using FFNNs. One of the advantages of using NNs is the capability to discover patterns in the data itself, which can help in finding the relationship between two different stocks (stock pairs). NNs have non-linear, non-parametric adaptive learning properties and have the most desirable outcome in modeling and forecasting. However, NNs have their drawbacks, such as the "over-training" problem, where the network loses its generalizability. The generalization capability of NNs is important when forecasting future stock prices (Lawrence, 1997). NNs can be applied to forecast price changes before divergence and after convergence periods. This will help prepare traders to take the correct trading positions (sell-short, buy-long).

Rule discovery can be considered a method of finding relationships between stocks or markets by studying the correlation between individuals (antecedent and consequent) (Thuraisingham, 1998; Han and Kamber, 2006). Association rule mining techniques are not specifically used to solve the pairs trading problem, but rather to provide traders with greater insight. For example, mining frequent two itemsets (two stocks) from stock data is a method of generating stocks rules. For example, "if IBM's stock price increases, the MSFT stock price is likely to increase too" and vice versa. Association rules can be used to predict the movement of the stock prices, based on the recorded data (Lu et al., 1998; Ellatif, 2007). This will help in finding the convergence in stock prices. However, association rule mining techniques usually generate a large number of rules; this presents a major interpretation challenge for investors.

Clustering can be considered the most important unsupervised learning technique in both the data mining and machine learning areas. Basalto et al. (2004) have applied a nonparametric clustering method to search for correlations between stocks in the market. This method, the Chaotic Map Clustering (CMC), does not depend on prior knowledge about a cluster, making it an optimal strategy to find pairs; originally proposed in (Angelini et al., 2000), it identifies similar temporal behavior of traded stock prices.

Lin et al. (2004) have used a genetic algorithm (GA) technique which has been applied to many financial problems (Chen, 2002; Allen and Karjalainen, 1999) to tackle the parameters problem in the trading process. A solution has been obtained by using a sub-domain for each parameter instead of one value. Lin et al. (2005) have subsequently applied the GAs to reduce the effect of noise in the input data. This noise can cause the system to generate unwanted alerts, which can mislead traders into making the wrong decisions.

The trading process is based on many rules that depend on many parameters. Trading rules help traders to decide what position to take regarding their shares (sell or buy). GA was used to find the best combination of parameters which is the first step for two other GA approaches presented in (Lin et al., 2005; Cao et al., 2006a).

Cao et al. (2006a) proposed a technique that has been used in mining stock pairs. In their approach, they used genetic algorithms combined with fuzzy operations. Fuzzy logic (Zadeh, 1965) was combined with (GAs) (Coley, 1999) because of the challenges that GAs encounter when dealing with domain-oriented businesses that consist of multiple-user requirements and demands. They used correlation to analyze the pairs' relationship by considering their correlation coefficient to find highly correlated stock in Australian Stock Exchange (ASX). As a result, they found unexpected pairs that are distant from the traders' expectations, and that most of the correlated stocks belonged to different sectors.

Cao et al. (2006b) introduced fuzzy genetic algorithms to mine pairs relationships and proposed strategies for the fuzzy aggregation and ranking to generate the optimal pairs for the decision-making process. They categorized the pairs into two classes: pairs that come from the same class, named "kindred", and others, named "alien". They classified the type of relationship between the pairs. The first type is the "negative relationship", where pairs are dissimilar (i.e., they move in opposite directions). The second type is the "positive relationship", where pairs follow a similar pattern. This classification of the pairs helps when using correlation and association mining techniques to predict decisions (Kovalerchuk and Vityaev, 2000; Chatfield, 2004).

The focus of these researchers (Lin et al., 2004, 2005; Cao et al., 2006a,b) who used GAs or a combination of GAs and fuzzy logic was to overcome the parameter obstacle. Therefore, they managed to use sub-range values for each of the parameters instead of using a single value. This ensures that the process of finding relationships between assets (stocks, rules) comes from optimal values that help in obtaining optimum pairs.

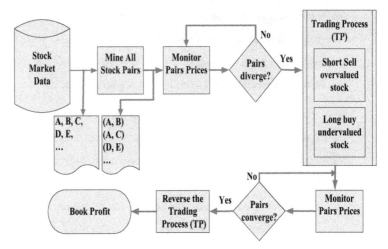

Figure 6.1: The complete Pairs trading process, starting from surfing the stocks data until making the profit.

Since none of these methods has used the DTW as the similarity measure to capture the shape of the stock sequences, a clear distinction between their approach and ours.

6.3   Pairs Trading Framework

This section describes the general framework of the pairs trading strategy. Figure 6.1 is a flowchart showing the five main stages of pairs trading.

1. **Mining stock pairs:** Any two stock prices that move together over time (i.e., in similar patterns). The simplest way of finding all pairs in a given stock dataset is by screening the entire dataset and returning stocks that form pairs with each other. This stage is important, because it identifies pairs that should be monitored. Our technique *SparseDTW* is used to report all pairs efficiently.

2. **Monitoring the spread:** An alert system is used to notify traders if there is a

change in the price in comparison with the pair's historical price series. Alerts will be sent if there is a noticeable divergence in the price (lose in one and win in another).

3. **Trading process:** An investor chooses the appropriate positions in the market. Once the divergence alerts are received, investors can decide how many shares they should buy or sell. Traders will determine the stocks for which they should take a short or long position. The main hope is that the stock prices will soon revert to their normal price levels.

4. **Looking for convergence:** This stage is similar to stage two, but convergence is monitored. When the pair's price series starts heading towards the normal historical price level, alerts will be generated to notify investors of the best time to gain a profit.

5. **Reverse stage 3:** After receiving alerts that show the pair's status is returning to normal, investors need to reverse the positions they initially took when the prices diverged.

This research used the ASX (S&P 50) indices instead of dealing with huge number of stocks. Table 6.2 lists the 50 indices and their codes.

To summarize the pairs trading strategy, Figure 6.2 illustrates an example of a simple index pair *Developer & Contractors* (XDC) and *Transport* (XTP) over 500 days (almost two years of price movements).

This example shows two price series moving together over time, which means that index XDC and XTP are a pair. After finding that indices XDC and XTP form a pair, their price series will be monitored. It is clear that prices start to diverge from day 200. Around day 240, prices show the greatest deviation. Then alerts will be issued to investors so they can

Code	Index Description	Code	Index Description
XAI	All industrials	XME	Media
XAM	All mining	XMI	Miscellaneous industrials
XAO	All ordinaries	XMJ	Materials
XAR	All resources	XNJ	Industrial
XAT	Alcohol & tobacco	XOM	Other Metals
XBF	Banks & finance	XPJ	Property trusts
XBM	Building materials	XPP	Paper & Packaging
XCE	Chemicals	XPT	Property Trusts
XDC	Developers & contractors	XRE	Retail
XDI	Diversified industrials	XSJ	Consumer Staples
XDJ	Consumer Discretionary	XSO	Small Ordinaries
XDR	Diversified resources	XTE	Telecommunications
XEG	Engineering	XTJ	Telecommunications Services
XEJ	Energy	XTL	20 leaders
XFH	Food & household goods	XTO	ASX 100
XFJ	Financials	XTP	Transport
XFL	50 leaders	XTU	Tourism & Leisure
XGO	Gold	XUJ	Utilities
XHJ	Health Care	XXJ	Financial & Property Trusts
XIF	Investment & financial services	XUI	Infrastructure & Utilities
XIJ	Information Technology	XAG	ASX/Russell all Growth
XIN	Insurance	XCN	Asian Listing
XJO	ASX 200	XTT	Trans-Tasman 100
XKO	ASX 300	XSF	Soild Fuel
XMD	MIDCAP 50	XOG	Oil & Gas

Table 6.2: Indices (S&P 50) in the Australian Stock eXchange (ASX).

take the right position in the market. In this example, investors should short sell index XDC (sell the winner) and long buy index XTP (buy the loser). Investors must continue to monitor the pair's price series to maximize their profit. From day 250, prices start reverting to their normal levels. Alerts will be generated again, this time showing that prices are normal and that investors can reverse their trades to make the maximum profit.

6.4   Finding Pairs Trading Approach

The previous sections provided an introduction and overview of the pairs trading strategy. In this section we show that our approach uses (*SparseDTW*) to mine all pairs.

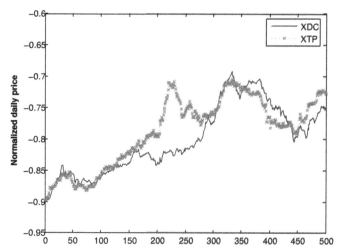

Figure 6.2: An example of pair of indices, *Developer & Contractors* (XDC) and *Transport* (XTP). Divergence in prices is clearly shown during the period of 200 days and 300 days.

6.4.1   Preprocessing Stock Data

The stock data used in this research comprises the daily stocks/indices prices. Our concern here is the stock prices that have been traded (the ask price matches with the bid). This research used the ASX data obtained in the methods described in Appendix C.

Stock prices are normally extracted from the order book[4]. This data has to be preprocessed before using it. The following steps show the preprocessing stage.

- Given the focus was on the price of the stocks, other attributes like order type and volume were removed.

- The stock prices have to be normalized (by locating all prices in a particular unit). The underlying reason is that each stock has its own unit, and this may lead to an unfair comparison. However, by normalizing the prices by the standard deviation

---

[4]The order book is a register (database) in a securities exchange that stores all public orders (sell and buy orders) as well as all transactions that are executed (traded).

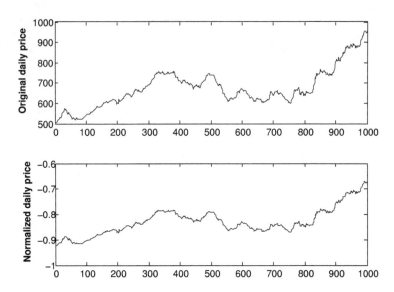

Figure 6.3: Index *All Industrials* (XAI) before and after the normalization.

the comparison between any two stocks will make more sense. The normalization
is performed using Equation 6.1.

$$nP_i^t = \frac{P_i^t - \overline{P_i^t}}{\sigma_i} \tag{6.1}$$

Where nP_i^t is the normalized price of stock i at time t, P_i^t is the actual price for
stock i, and $\overline{P_i^t}$ is the average price of stock i. σ_i is the standard deviation of stock
i within a moving window.

Figure 6.3 plots the original and normalized daily prices of the index *All Industrials*
(XAI). This figure shows clearly that the price trends are very similar before and after the
normalization.

#	Pair		EucDist	DTW
1	XDC	XTP	3.39	0.73
2	XAI	XRE	2.01	0.39
3	XJO	XKO	6.33	0.13
4	XIN	XFH	11.4	0.49
5	XXT	XFJ	28.92	0.65

Table 6.3: Comparison between the EucDist and DTW measures for five different pairs.

6.4.2   Choosing a Proper Similarity Measure

Our focus is monitoring the similarity in the price movement. Therefore, similarity measures should be used to evaluate how the prices of particular pairs are close to one another.

Similarity or dissimilarity can be clearly captured using a distance measure. The issue here is to choose the proper (dis)similarity measure. Two of the most famous measures in the area of time series are the Euclidean (EucDist) and Dynamic Time Warping (DTW) distances. An overview about these two measures is given in Chapter 2 (Section 2.6). Since EucDist has several limitations (as discussed in Chapter 2), DTW is used as the similarity measure here. Table 6.3 shows the distances between five different pairs, using EucDist and DTW measures. If we are using a similarity threshold = 1, using EucDist will lead to missing all of these pairs.

6.4.3   Report Stock Pairs

We describe the process of finding stocks pairs using our SparseDTW as the similarity mining technique. This is performed through the following steps.

- After normalizing the stocks/indices prices, DTW distance is used to check stocks that behave similarly; in other words, we find stock pairs that have their DTW distance is the minimum.

Algorithm 6.1 The sequential search algorithm pseudocode.

Input: Q: Query sequence of length n, Time Series Database TSD.

1: $BestSoFar = \infty$
2: **for all** $C \in TSD$ **do**
3:     $LB\_Dist = LowerBoundDistance(C_i, Q)$
4:     **if** $LB\_Dist < BestSoFAR$ **then**
5:         $True\_Dist = \textbf{SparseDTW}(C_i, Q)$
6:         **if** $True\_Dist < BestSoFar$ **then**
7:             $BestSoFar = True\_Dist$
8:             $IndexOfBestMatch = i$
9:         **end if**
10:     **end if**
11: **end for**

#	Pair		DTW	SparseDTW
1	XDC	XTP	0.73	0.73
2	XAI	XRE	0.39	0.39
3	XJO	XKO	0.13	0.13
4	XIN	XFH	0.49	0.49
5	XXT	XFJ	0.65	0.65

Table 6.4: Comparison between standard DTW and SparseDTW techniques when calculating the DTW distance for five different pairs.

- The DTW distance is computed using SparseDTW, which always returns the optimal distance between any two time series (stocks).

Figure 6.4 shows the dendrogram of the ASX indices after clustering them using hierarchical clustering. We obtained the DTW distance matrix using the *SparseDTW* algorithm. Three of the top closest pairs were reported after using a distance threshold of 1; the dashed line indicates the threshold. Table 6.4 provides a comparison between the standard DTW and the SparseDTW techniques that used in calculating the DTW distance between indices. The table clearly shows that SparseDTW gives the exact distance as that obtained from standard DTW.

- Usually, time series datasets contain huge number of stocks; hence, we need an efficient technique that scans through the entire dataset to find all pairs. Therefore, we have used a sequential search technique which incorporates a lower bound

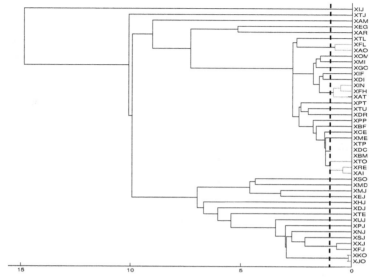

Figure 6.4: Dendrogram plot after clustering the ASX indices. The closest three pairs (smallest distance between the indices) are chosen as examples of pairs trading pattern.

technique that was proposed by Keogh and Ratanamahatana (2004) to reduce the number of *SparseDTW* computations, thus speeding up the sequential scan search for the query stock. Algorithm 6.1 shows the pseudocode for the sequential search technique. This finds the index of the closest candidate stock C to the query stock Q, and it calculates the DTW distance between them only if the lower bound distance satisfies the condition $LB\_Keogh(Q, C) \leq BestSoFar$, where $BestSoFar$ is treated as predefined threshold. The idea is to use a cheap lower bounding calculation which will allow us to perform the expensive calculations when are they critical. Equation 6.2 shows the lower bound distance between stocks Q and C. To give an example about the way the lower bound LB_Keogh works, Figure 6.5 depicts two sequences – index *ASX 200* as the query sequence, and *ASX 300* as the candidate sequence – and the lower and upper bounds (U and L, respectively) of the query sequence.

$$LB\_Keogh(Q,C) = \sqrt{\sum_{i=1}^{n} \begin{cases} (c_i - U_i)^2 & \text{if } c_i > U_i \\ (c_i - L_i)^2 & \text{if } c_i < L_i \\ 0 & otherwise. \end{cases}} \qquad (6.2)$$

LB_Keogh uses a refine-and-filter strategy, where U and L are used as an envelope around the query sequence. This is similar in behavior to the minimum bounding rectangle (MBR) strategy.

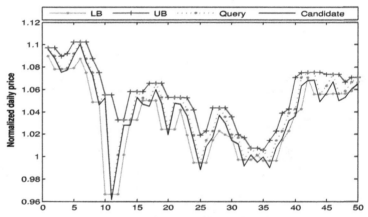

Figure 6.5: Plot of index *ASX 200* as a query sequence and *ASX 300* as a candidate sequence from the ASX index daily data. LB and UB are the lower bound and upper bound of index *ASX 200*, respectively.

- After reporting all pairs, it is necessary to monitor them and look for signals which indicate that the proper time for taking the suitable position in the market. This can be performed by applying several trading rules. Examples of such rules are described in Section 6.4.4.

Figure 6.6(a) and Figure 6.6(b) are two examples of index pairs in the ASX. Figure 6.6(a) shows two indices which behave for most of the time similarly. Figure 6.6(b) shows example of an identical pair.

An example of two indices which behave differently from each other is given in Figure 6.7. The *Information Technology* (XIJ) and *Retail* (XRE) are from two different industries. Investors can easily understand that the price movement in XIJ will not be influenced by XRE and vice versa. Therefore, looking for pairs trading patterns in these two indices is meaningless.

The number of reported pairs depends on the sectors where the stocks belong. Two major features of the sectors are as follows:

1. The sector volatility: If the sector is showing high volatility, it will produce few pairs.

2. The sector homogeneity: Since the homogeneity can be a feature of similarity, we know that pairs should be very similar; therefore, sectors with high homogeneity should produce large number of pairs.

For example, a commercial services sector is expected to have few pairs; on the other hand a financial sector should give huge number of pairs because of the enormous number of trades and homogenous operations between companies.

6.4.4 Trading Process (TP)

After we have shown the process of reporting the pairs from stock datasets, those pairs can be used for taking position (short sell or long buy) in the market. This should not be performed arbitrary, since the period of the appropriate position is an important concern.

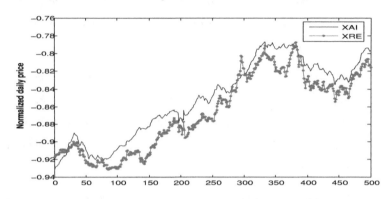

(a) An example of pair of indices, *All Industrials* (XAI) and *Retail* (XRE).

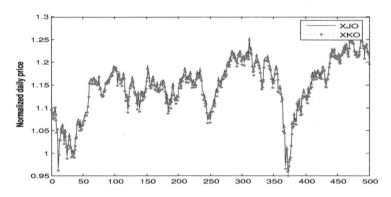

(b) An example of an identical pair of indices, *ASX 200* (XJO) and *ASX 300* (XKO).

Figure 6.6: Two examples of index pairs.

Therefore, we will describe several trading rules which can decide the suitable time for taking a position (i.e., the proper time for opening and closing a trade. Two of the trading rules are as follows:

1. **Mean and two-standard-deviations rule:**

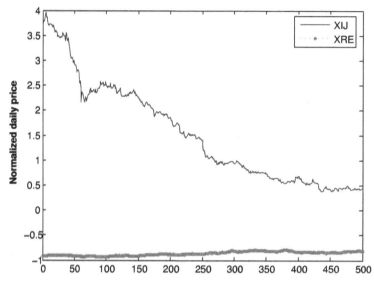

Figure 6.7: An example of two indices, *Information Technology* (XIJ) and *Retail* (XRE), which do not form pair.

This rule was proposed by Herlemont (2004) to open a position when the ratio of two stock/index prices hits the two rolling standard deviation and to close it when the ratio returns to the mean. Generally, the position should not be opened when the ratio breaks the two-standard-deviations limit for the first time, but rather when it crosses it to reach to the mean again. This will avoid opening a position in a pair with a wide spread. Figure 6.8 shows the entry/open and the closing positions, as well as the limits (mean and two-standard-deviations).

2. **Absolute distance rule:**

This uses the absolute distance between the pair prices (Equation 6.3). The positions are taken on the basis of the value of this distance; if the absolute distance $AbsDist > t$, where t is a predefined threshold, then a *short sell* position will be appropriate on the overvalued stock and a long buy on the undervalued stock.

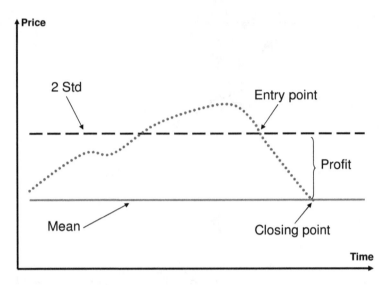

Figure 6.8: Pairs Trading rules.

$$AbsDist = |nP_i^t - nP_j^t|, \tag{6.3}$$

where nP_i^t and nP_j^t are the normalized prices at time t of stock i and j, respectively.

6.5 Summary and Conclusions

Pairs trading is an investment strategy that depends on the price divergence between a pair of stocks. Essentially, the strategy involves choosing a pair of stocks that historically move together, taking a long-short position if the prices of the pair diverge and reversing the previous position when those prices converge. The rationales of pairs trading is to make a profit and to avoid market risk. We have shown in this chapter a description of state-of-the-art research conducted in finding pairs trading patterns. Since pairs trading is useful to many investors, we have discussed the use of a similarity measure (DTW) when finding stocks pairs. DTW is a superior similarity measure because of its ability to capture the stock sequences.

We have shown the use of our algorithm SparseDTW, described in Chapter 5, in reporting all stock pairs from large stock-market data. SparseDTW can be combined with lower bounding techniques that increase the search technique by reducing the number of SparseDTW computations. The datasets used in this chapter are obtained from the ASX on the basis of their suitability for our approach. The preparation process is set out in Appendix C.

Chapter 7

Conclusion and Future Work

This chapter presents a summary and the implications of this research over different domains. It provides additional research directions that could be extended in the future.

7.1   Summary of the Research

Data mining refers to the extraction of knowledge from large amounts of data. It is an analysis tool which can be combined with sophisticated mechanisms to analyze large datasets. Data mining incorporates a variety of different techniques that aim to discover diverse types of patterns from a given database, based on the requirements of the domain. These techniques include association rules mining, classification, cluster analysis and outlier detection.

Due to the wide-range of applications that deal with spatial, spatio-temporal (ST) and time series data (TSD), we have conducted the research reported in this book. The research was in four parts, each of which addressed a different problem.

7.1.1 Mining Complex Co-location Rules

The widespread use of applications that produce large amounts of spatial data has motivated many researchers to propose new tools for mining such data. The purpose of these
tools is to discover interesting knowledge that can contribute to the application domain by
adding new facts and proving existing facts. Many researchers have proposed interesting
tools to mine patterns that are hidden in large spatial databases. Several studies targeted
one type of spatial pattern, known as a "co-location pattern" – a group of objects, such as
galaxies, in which each object is located within a given distance of another object in the
group. Mining spatial co-location patterns is an important spatial data-mining task with
wide-ranging applications in areas such as public health, environmental management and
ecology (Huang et al., 2004). Co-location rules are defined as signs that specify the presence or absence of spatial features in the vicinity of other spatial objects. Examples of
such rules are given with more explanation in Chapter 3.

To mine such rules, the spatial data must first be transformed into a transactional-type
dataset, to allow the association rule mining technique to be applied. The transformation
is achieved by extracting co-location patterns (maximal clique patterns) from the raw
spatial data. A maximal clique pattern is one that does not appear as a subset of another
clique in the same co-location pattern. We focused on using maximal cliques to allow the
mining of interesting *complex spatial relationships* between the object types as discussed
in Chapter 3.

We have proposed an efficient algorithm (GridClique) that extracts maximal clique patterns from a large spatial dataset (astronomy dataset). The GridClique algorithm is based
on a divide-and-conquer strategy that divides the space into grid structure and finds all
neighbors of each spatial object. It checks all neighbors against the co-location condi-
tions (i.e., that each object must be within a given distance of another). The extracted

maximal clique patterns are used as raw data to the association rule mining technique, where the maximal clique's ID is considered as the transaction's ID, and the maximal clique's members are the transaction items.

Previous research has not used the concept of *maximal clique* when mining complex co-location rules. Our experiments, which were carried out on a real-life dataset obtained from an astronomical source – the Sloan Digital Sky Survey (SDSS) – show that our results are potentially valuable to the field of astronomy and can be interpreted and compared easily in relation to existing knowledge.

7.1.2   Mining Complex Spatio-Temporal Patterns

The extensive availability of location-aware devices (such as GPS devices) promotes the capture of the detailed movement trajectories of people, animals, vehicles and other moving objects, opening new opportunities for better understanding the processes involved. The data generated is ST data, which contains the spatial evolution of objects over time.

Data mining research focuses on developing techniques to discover new ST patterns in large repositories of ST data. We have focused on discovering fixed-subset flock patterns (where objects are moving close together in coordination). Benkert et al. (2006) described efficient approximation algorithms for reporting and detecting flocks. Their main approach has an exponential dependency on the duration of the flock pattern. This motivated us to propose an approach that reduces this dependency. To accomplish this, we have combined their approximation logarithm that reports flock patterns with a dimensionality reduction technique (random projections). The random projections technique generates a sparse random matrix that is used to project the original data into lower-dimensional space, as described in Chapter 4. To the best of our knowledge, this is the first time random projections have been used to reduce dimensionality in the ST setting presented in this book. We have proved that random projections will return the "correct" answer with high probability. Our experiments on real-life, quasi-synthetic and synthetic datasets strongly support our theoretical bounds.

7.1.3 Mining Large Time Series Data

Many applications, such as those in computational biology and economics, produce TSD. The data generated by those applications continually grows in size, and places an increased demand for developing tools that capture the similarities among them. An interesting example of a real-life query, which can be answered by reporting the similarity between sequences, is financial sequence matching, where investors plan to monitor the movement of stock prices to obtain information about price-changing patterns or stocks that have similar movement patterns. An example of such patterns in this sector is called "pairs trading", where investors search for knowledge to increase expected profit and reduce expected loss.

Dynamic time warping (DTW) is a distance measure that allows sequences, which are of different lengths, to be stretched along the time axis to minimize the distance between them. Since TSD usually consists of sequences of different lengths, DTW has been used as the underlying similarity measure in this research. Although DTW is a better similarity measure, it has expensive space complexity, which leads to long computation times. This has prompted many researchers to develop several techniques to increase its speed by reducing the search space.

We have devised an efficient algorithm (SparseDTW) which exploits the possible existence of inherent similarity and correlation between the two time series whose DTW is being computed. We always represent the warping matrix using sparse matrices, which lead to better average space complexity than other approaches. The SparseDTW technique can easily be used in conjunction with lower bounding approaches. We believe that no previous technique has used the idea of sparse matrices to reduce the search space. Our experiments show that SparseDTW gives exact results compared with other techniques which give approximate or non optimal results.

7.1.4 Mining Pairs Trading Patterns

Many algorithms have focused on discovering useful patterns in stock market datasets. One of the valuable patterns is called pairs trading; that is, a strategy that involves buying the undervalued stock and short-selling the overvalued one, thus maintaining market neutrality. Investors need to watch stock price movements (changes) to find interesting patterns that indicate an increased chance of making a profit. Reporting stocks (stock pairs), that are correlated or that share a high level of similarity with each other, is a very challenging problem.

This book has used DTW as the similarity measure due to its capability of capturing the sequences' shape. Previous approaches have never used DTW to mine similarities between stock market data – specifically when mining pairs trading patterns. We believe that we have achieved our main goal of accurately reporting all pairs trading patterns from large daily TSD, such as stock market data. We have proposed a framework to find pairs trading patterns in large stock market data efficiently. We have successfully applied our proposed algorithm (SparseDTW), which we proposed to speed DTW, to report all stocks pairs. Our experiments show that SparseDTW is a robust tool for mining pairs trading patterns in TSD.

7.2 Implications to Different Domains

This section describes the implications of the proposed approaches for different domains.

- A new algorithm (GridClique) is proposed for extracting complex co-location patterns from large spatial dataset (astronomy dataset). This algorithm is the first of its nature because it extracts complex maximal clique patterns to be used in mining complex co-location rules. The GridClique algorithm has many potential uses in diverse application areas where complex co-location rules may provide valuable insights. Examples of these applications include: spatial data, climate and bioinformatics analyses. In bioinformatics, the GridClique algorithm can be applied on the Electron Microscopy Data Bank (EMDB) data to discover relationships between nucleic acids and protein sequences.

- Another approach has been proposed to tackle the dimensionality curse in large ST setting using the random projections technique. We have shown, for the first time, that this approach produces "correct" results when extracting long-duration flock patterns. The proposed approach which is used to mine long duration flock patterns can be applied in many applications, such as moving object surveillance, financial and ST applications.

- The SparseDTW algorithm as a technique to increase the speed of DTW by reducing its space complexity. The core idea of this algorithm is to dynamically exploit the similarity and correlation between the two time series whose DTW is being computed. This idea is the first of its kind that used to speed up DTW. In addition to the use of SparseDTW in mining pairs trading patterns, it can be used in many other applications that generate TSD, such as networks fault management and speech recognition.

7.3 Future Work

This book has proposed several approaches to solve significant problems, suggesting new research directions in three areas: spatial data mining (1 and 2), ST data mining (3 and 4), and TSD mining (5 and 6).

1. The GridClique algorithm can be modified to mine patterns such as stars and other generic co-location patterns. This may allow for a different interpretation of the complex co-location rules being mined.

2. One of the valuable directions to apply the GridClique algorithm is on the Electron Microscopy Data Bank (EMDB) data, which contains experimentally determined three-dimensional maps. The bank contains biological data including nucleic acid, protein sequences and macromolecular structures. This could help scientists to understand the relationship between protein sequences.

3. Our approach can be extended from mining long-duration flock patterns to mine other patterns, such as convergence and encounter in large real-life ST data.

4. An interesting new direction of the flock mining approach is the use of association rule mining type algorithms to discover novel flock-like patterns. For example, it is known that if F is an (m, k, r)-flock then every subset of F of size $m' < m$ is an (m', k, r)-flock.

5. SparseDTW can be used for detecting outliers in large TSD, where the outliers are defined as those patterns that have their DTW distance larger than a given threshold.

6. A valuable research direction is to extend SparseDTW for streaming setting. The aim would be to monitor TSD looking for predefined patterns. For example, using SparseDTW for network fault management purposes or in stock market data where noticeable price changes need to be reported.

Appendix A

Spatial Data Preparation

his appendix provides a comprehensive details about the data preparation steps used to prepare the astronomy data – Sloan Digital Sky Survey (SDSS). This data has been used when mining complex co-location rules (Chapter 3). Table A.1 provides the notations used in this appendix.

A.1 Data Extraction

This section describes the method for extracting attributes from the SDSS Data Release 6 and using them to categorize galaxy objects. A view called *SpecPhoto* which is derived from a table called *SpecPhotoAll* is used. The latter is a join between the *PhotoObjAll* and *SpecObjAll* tables. In other words, *SpecPhoto* is view of joined *Spectro* and *PhotoObjects* which contains only the clean spectra.

The concern is to extract only the galaxy objects from the SDSS using the parameter (object type=0). The total number of galaxy type objects stored in the SDSS catalogue is $507,594$. To ensure the accuracy for calculating the distance between objects and the

Symbol	Description
parsec	Unit of length used in astronomy. It stands for "**par**allax of one arc **sec**ond"
Mpc	An abbreviation of "Mega-parsec", which is one million parsecs, or 3261564 light years
arcmin	Unit of angular measurement. Sizes of objects on the sky, field of view of telescopes, or practically any angular distance "Arc of Minutes"
z	RedShift
zWarning	Parameter used to guarantee that the corrected RedShift values are used
zConf	RedShift confidence
X	X-coordinate
Y	Y-coordinate
Z	Z-coordinate
SDSS	Sloan Digital Sky Survey
U	Ultraviolet
R	Red light magnitude
r-band	r-band Petrosian magnitude
H_o	Hubble's constant
LRG	Luminous Red Galaxies

Table A.1: Description of the notations used.

No	Field name	Field description
1.	specObjID	Unique ID
2.	z	Final RedShift
3.	ra	Right ascention
4.	dec	Declination
5.	cx	x of Normal unit vector
6.	cy	y of Normal unit vector
7.	cz	z of Normal unit vector
8.	primTarget	prime target categories
9.	objType	object type : Galaxy =0
10.	modelMag_u	Ultraviolet magnitude
11.	modelMag_r	Red Light magnitude

Table A.2: The SDSS schema used in this work.

earth, which leads to calculate the X, Y, and Z coordinates for each galaxy, number of parameters are used, such as $zConf < 0.95$ (the rigid objects) and $zWarning = 0$ (correct RedShift). Therefore, the number of galaxy objects is reduced to $365, 425$.

SDSS release DR6 provides a table called *Neighbors*. This table contains all objects

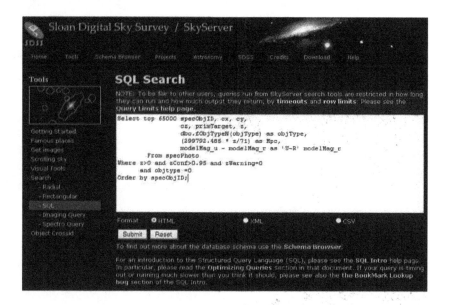

Figure A.1: The front end of the SDSS SkyServer. It provides an SQL search facilities to the entire stored data.

that are located within 0.5 arcmins, this makes it not useful in this research because there is no ability to choose any other distance greater than 0.5 arcmins to form the neighborhood relationship between objects. For example, in our experiments $(1, \ldots, 5)$ Mpc[1] (distances) are used as the thresholds to check if objects are co-located or not. Table A.2 lists the extracted fields from the SDSS (DR6) that have been used during the preparation process.

The raw data was obtained from SDSS (DR6) (Survey, 2005). This data is extracted from the online catalogue services using several SQL statements. The catalogue offers a very elegant interface that allows users to extract easily the preferred data (Figure A.1). The catalogue provides other tools that can be used to browse all tables and views in the

---

[1] See http://csep10.phys.utk.edu/astr162/lect/distances/distscales.html for details.

SDSS data. This catalogue is accessible from the SDSS website[2].

The first SQL statement used to extract the first 65000 objects (galaxies) is as follows:

```
Select top 65000 specObjID, cx, cy,
                 cz, primTarget, z,
                 dbo.fObjTypeN(objType) as objType,
                 (299792.485 * z/71) as Mpc,
                 modelMag_u - modelMag_r as 'U-R' modelMag_r
        From specPhoto
Where z>0 and zConf>0.95 and zWarning=0
      and objtype =0
Order by specObjID;
```

The second SQL statement used to extract the second 65000 objects starting from the last object's ID (119286486972497000) is as follows:

```
Select top 65000 specObjID, cx, cy,
                 cz, primTarget, z,
                 dbo.fObjTypeN(objType) as objType,
                 (299792.485 * z/71) as Mpc,
                 modelMag_u - modelMag_r as 'U-R' modelMag_r
        From specPhoto
Where z>0 and zConf>0.95 and zWarning=0
      and objtype =0
      and specObjID>119286486972497000
Order by specObjID;
```

---

[2]http://cas.sdss.org/dr6/en/tools/search/sql.asp

The second SQL statement is different from the first one by adding the a condition $specObjID > \{lastID\}$. The reason behind extracting just 65000 objects is to be able to handle them using Microsoft Excel 2003, which was used to cleaning to some records.

A.2 Data Transformation

The extracted data needs to be transformed into the right format before start mining it. Transforming the data makes it accessible using $Oracle10_g$, where we uploaded the data into a normalized database. The use of the Oracle helps us in: (i) manipulating the extracted data and report some statistics about it, (ii) eliminating the undesired fields that we had to extract when we initially downloaded the data from the SDSS repository and (iii) calculating the distance between galaxy objects and the earth. Few tables were created to store the extracted data. We created number of stored procedures to categorize galaxy objects and put the data into the right format.

A.3 New Attributes Creation

Having all necessary fields extracted, the next step is to calculate for each galaxy the exact value of the X, Y, and Z coordinates that are not stored in the SDSS data. The following steps show the process of calculating the coordinates of each galaxy:

1. Calculating the distance between the earth and galaxy objects using Hubble's law and redshift[3] z value (Equation A.1).

$$D \approx \frac{c \times z}{H_o}. \tag{A.1}$$

---

[3]is a shift in the frequency of a photon toward lower energy, or longer wavelength.

Where c is the speed of light, z is the object RedShift, and H_o is Hubble's' constant. Currently the best estimate for this constant is 71 $kms^{-1}Mpc^{-1}$ (Spergel et al., 1997; M. and Churchman, 2005).

2. Considering the extracted unit vectors cx, cy, and cz and multiplying them by D that obtained form the previous step. Equations A.2, A.3 and A.4 used to calculate the final value of X, Y and Z coordinates, respectively.

$$X = D \times cx. \tag{A.2}$$

$$Y = D \times cy. \tag{A.3}$$

$$Z = D \times cz. \tag{A.4}$$

A.4   Galaxies Categorization

Our purpose is to mine complex co-location rules in astronomy data. Therefore, we have to add to our prepared data the galaxy types. More specifically, to use the association rule mining technique on the prepared data the galaxy type is used. If we go back to the supermarket example $\{bread \rightarrow cheese\}$ rule, instead we use, for example, $\{Galaxy\ type\ A \rightarrow Galaxy\ type\ B\}$. This rule can be interpreted as the presence of galaxy type A implies the presence of galaxy type B.

Galaxy types were not existed in the SDSS data. Therefore, we used several parameters to find the galaxy objects types. That is performed as follows:

Object ID	Object type	X-Coordinate	Y-Coordinate	Z-Coordinate
1	LRG-Late	2.5	4.5	5.5
2	Main-Early	6	4	2.5
3	LRG-Early	2	9	11
4	Main-Late	1.5	3.5	3
5	LRG-Late	5	3	4.5
6	Main-Early	7	1.5	2.5

Table A.3: An example to show the final data after the preparation.

1. The difference between Ultraviolet U and Red light magnitude R, is used to categorize galaxy objects into either "Early" or "Late". If the difference ≥ 2.22 the galaxy is considered to be "Early", otherwise "Late".

2. The value of the r-band *Petrosian* magnitude indicates whether a galaxy is "Main" (close to the earth) or "Luminous Red Galaxies" LRG (far away from the earth). If r-*band* ≤ 17.77 the galaxy object is "Main", otherwise it is LRG (Martin and Saar, 2002).

Consequently, four galaxy types were found as a combination of the above mentioned types. These categories are: **Main-Late, Main-Early, LRG-Late**, and **LRG-Early**.

A.5   Final Format of the Spatial Data

Table A.3 displays an example of the final format of the SDSS data after the pre-processing process. This final format is used when mining maximal clique patterns, which are the co-location patterns.

A.6 Summary

We have provided in this appendix the preparation process that was performed on the spatial data that was extracted from the SDSS catalogue.

Appendix B

Spatio-Temporal Data Preparation

This appendix describes the pre-processing step (i.e. data preparation) applied on the raw datasets that used in Chapter 4. The data was used for mining long duration flock patterns. Table B.1 lists all notations used in this appendix.

B.1 Spatio-Temporal (ST) data

We used both real-world datasets and synthetic datasets to demonstrate the efficiency of our proposed approach. Table B.2 gives an example of the final format of the used ST data.

B.1.1 Synthetic datasets

Twenty datasets with varying number of points, number of flocks and duration of flocks were created. In particular, five datasets each of size 16K, 20K, 32K, 64K and 100K were seeded with 32, 40, 64, 128 and 200 flocks respectively of duration (number of time step) 8, 16, 500 and 1000. The size of each flock was set to 50 entities and the radius was fixed

Symbol	Description
NDSSL	Network Dynamics and Simulation Science Laboratory
RP	Random Projection
PCA	Principle Component Analysis
ST	Spatio-Temporal
$X_{n \times d}$	Original dataset
$R_{n \times \kappa}$	Random matrix
$rand(\cdot)$	Function to generate random numbers between 0 and 1
$princomp(\cdot)$	Function to retrieve the principle Components of a matrix

Table B.1: Description of the notations used.

Object	Location 1		Location 2		$\cdots$	Location 359	
ID	X	Y	X	Y	$\cdots$	X	Y
13313	65.106	-139.317	65.228	-139.44	$\cdots$	65.739	-139.481
13314	66.45	-139.169	66.453	-139.174	$\cdots$	66.473	-139.254
17733	66.145	-139.668	67.143	-138.972	$\cdots$	67.381	-138.951
17734	67.29	-137.994	67.542	-137.699	$\cdots$	67.586	-137.592
17735	69.559	-141.56	69.571	-140.442	$\cdots$	68.352	-138.041
18750	67.395	-138.79	67.429	-138.835	$\cdots$	67.663	-138.471
18757	66.286	-138.502	66.535	-138.527	$\cdots$	66.678	-138.833
18762	69.555	-141.49	69.538	-141.133	$\cdots$	69.503	-139.771

Table B.2: An example of 8 caribou cow locations. This data is an example of the ST data that we used in this book.

to 50 units. In the original data (before the random projection), each point coordinate was selected from the integer interval $[0, \dots, 2^{16}]$.

B.1.2 Real-world datasets

Three different real-world datasets were used in our experiments. These datasets were generated by three different projects, namely, Network Dynamics and Simulation Science Laboratory (NDSSL) (pan, 2006), Caribou Herd Satellite Collar Project (pch, 2007), and Mobile Users profiles (Taheri and Zomaya, 2005).

When preparing the caribou data, we faced a problem, that is, trajectories are not of the same length. Therefore, we had to adjust the length of the short trajectories by repeating the last location. Program B.1 displays the code used to put the Caribou data in the desired ST format.

Program B.1 Preparing the Caribou dataset.

```
 Data = Load unformatted caribou data.
 IDs  = All Caribou's ID numbers
for i = 1 to  //Number of IDs
   RawData=[];
   idx = find(Data(:,1)==IDs(i,1));
   NumTimeSteps= size(idx,1);
   RowData=[IDs(i,1)];
   for j = 1 : NumTimeSteps
      RawData = cat(2,RawData, Data(idx(j,1),2)); % Latitude
      RawData = cat(2,RawData, Data(idx(j,1),3)); % Longitude
   end
end
// Filling the gabs
n = Number of rows in RawData.
d = number of dimensions in RawData.
for i=1 to n
   Zeroes = find(Data(i,:) == 0);
   [r c] = size(Zeroes);  % The size of the Zeroes matrix
   NoRep = (col) - c;
   x= Data(i,NoRep - 1);
   y= Data(i,NoRep);
   for j=NoRep+1 to d step 2
       Data(i,j)   = x;
       Data(i,j+1) = y;
   end % inner for
end  % Outer for
Return( Data ) // final format
```

Program B.2 shows the technique that used to prepare the NDSSL data. NDSSL data contained originally a lot of missing locations in the middle of the trajectories. Therefore, the missing locations had to be filled before using it in our method. This was conducted in two steps:

1. Filling the missing locations in the middle of the trajectory: We assumed that the GPS was faulty and the object's location was not sent. Hence the last location was repeated to fill the missing location.

2. Making all trajectories of the same length: If a trajectory is short, we assumed that the object stopped before finishing its trip, or the GPS device stopped sending details because of some defects. To handle this we assumed that the object stopped and thus the last location is repeated until the trajectory length is equal to the longest trajectory.

B.2 Dimensionality Reduction Techniques

This section gives the pseudocode of the two dimensionality reduction techniques that have been applied on the ST datasets. Algorithms B.1 and B.2 show the pseudocode of RP and PCA techniques, respectively.

B.3 Summary

We have provided in this appendix the preparation process applied on the ST datasets as well as the pseudocodes of the two techniques used to tackle the curse of dimensionality.

Program B.2 Preparing the NDSSL dataset.

```
L= Locations;
A= Activities;
// get rid from the unwanted columns.
// We need Column #2 (Object ID) because it is unique,
// and Column#7.
// We need to find all locations
AA=cat(2,A(:,2),A(:,7));
m = max(AA(:,1));
// Obtain the maximum ID to be the limit of the iterations
// we need to find the location for each ID.
[rL cL]=size(L);
for i = 1 : m
    idx=find(AA(:,1)==i);
    [r c]=size(idx);
    tempLocation=[];
    tempLocation=i;
    for j=1 : r
        lID = AA(j,2);
        if (lID <= rL)
            X = round(L(lID,2));
            tempLocation=cat(2,tempLocation,X);
            Y = round(L(lID,3));
            tempLocation=cat(2,tempLocation,Y);
        end
    end
end
Return(tempLocation);
```

Algorithm B.1 The Random Projection pseudocode.

Input: $X_{n \times d}$: Original data.

  κ: number of desired dimensions.

Output: $E_{n \times \kappa}$: Projected data.

1: $X \Leftarrow$ Load the raw data.
2: $d \Leftarrow$ number of dimensions (X);
  {Generating random matrix R.}
3: $R \Leftarrow zeroes(d, k)$; Initialize R with zeroes.
4: $M \Leftarrow rand(d, k)$; This generates a matrix of entries between 0 and 1.
  {Filling R based on Achlioptas (2003)}.
5: $R(find(M > 5/6)) = \sqrt{(3)}$;
6: $R(find(M < 1/6)) = -\sqrt{(3)}$;
7: $E \Leftarrow X \times R$.
8: **return** E

Algorithm B.2 The Principle Component Analysis (PCA) pseudocode.

Input: $X_{n \times d}$: Original data.

  κ: number of desired dimensions.

Output: $E_{n \times \kappa}$: Projected data.

1: $X \Leftarrow$ Load the raw data.
2: $d \Leftarrow$ number of dimensions (X);
3: $SCORE \Leftarrow princomp(X)$;
  {Get the highest number of vectors based on the value of κ.}
4: $Vectors \Leftarrow SCORE(:, 1 : \kappa)$;
5: $E \Leftarrow Vectors$;
6: **return** E

Appendix C

Time Series Data Preparation

This appendix illustrates the pre-processing step (i.e. data preparation) applied on the time series datasets which were used in Chapter 5 and 6. Table C.1 lists the notations used in this appendix.

C.1   Time Series Data

In this book, we were exposed to different applications that produce time series data. Chapter 5 has given some information about some of these datasets. In this section we will categorize the time series datasets into two groups on the basis of their sources, namely, University of California - Riverside (UCR) and Australian Stock eXchange (ASX) datasets.

C.1.1   UCR Data

Keogh (2006) has made a collection of time series datasets. These datasets are publicly available from (Keogh, 2006). However, not all datasets were in the format that we are

Symbol	Description
SIRCA	Securities Industry Research Centre of Asia-Pacific
ASX	Australian Stock eXchange
UCR	University of California - Riverside

Table C.1: Description of the notations used.

using. Therefore, some techniques were used to put it in the desired format such as flipping the data from column data into row data in addition to deleting unwanted fields.

C.1.2   ASX Data

We managed to download the stock datasets from two different sources. The yahoo finance[1] and the Securities Industry Research Centre of Asia-Pacific (SIRCA)[2], which is a not-for-profit financial services research organization involving twenty-six collaborating universities across Australia and New Zealand. From both sources we downloaded the ASX data.

Program C.1 illustrates a script that used to download the stock/index data from yahoo finance.

C.2   Summary

We have provided in this appendix the preparation process applied on the time series datasets which used in mining complex time series patterns. We have categorized the datasets into two groups based on their sources, namely, UCR and ASX data.

---

[1] http://au.finance.yahoo.com/
[2] http://www.sirca.org.au/

Program C.1 AutoIt code to download stock market data.

```
; AutoIt version 3.0

;Run iexplorer.exe

Run("C:\Program Files\Internet Explorer\iexplore.exe")

Local $NumIndices = 50
;Declaring array

Dim $Indices[$NumIndices] = ["XAI","XAM","XBF",
                            "XCE","XRE", ....]
For $i = 0 to $NumIndices - 1
    $OutPut="C:\Documents and Settings\Ghazi\Desktop\d\"
    $OutPut&=$Indices[$i]
    $OutPut&=".csv"
    MsgBox(0, "We are downloading index=", $Indices[$i])
    $TheURL="http://ichart.finance.yahoo.com/table.csv?s="
    $TheURL&=$Indices[$i]
    $TheURL&="&a=00&b=1&c=2000&d=11&e=31&f=2006&g=d&ignore=.csv"
    InetGet($TheURL,$OutPut)
Next
```

Bibliography

Workshop on spatial data mining: Consolidation and renewed bearing. in conjunction with SIAM-DM, June 2006. URL http://ndssl.vbi.vt.edu/opendata/.

Network dynamics and simulation science laboratory. http://ndssl.vbi.vt.edu/opendata/index.html, July 2007. URL http://ndssl.vbi.vt.edu/opendata/index.html.

Porcupine caribou herd satellite collar project. http://www.taiga.net/satellite/, July 2007. URL http://www.taiga.net/satellite/.

The ireland story, November 2008. URL http://www.wesleyjohnston.com/users/ireland/map_index.html.

Wildlife tracking projects with GPS GSM collars. http://www.environmental-studies.de/projects/projects.html, August 2008. URL http://www.environmental-studies.de/projects/projects.html.

John Aach and George M. Church. Aligning gene expression time series with time warping algorithms. *Bioinformatics*, 17(6):495–508, 2001.

Tamas Abraham and John F. Roddick. Discovering meta-rules in mining temporal and spatio-temporal data. In *Proceedings of the Eighth International Database Workshop, Data Mining, Data Warehousing and Client/Server Databases (IDW)*, pages 30–41, Hong Kong, 1997. Springer-Verlag.

Tamas Abraham and John F. Roddick. Survey of spatio-temporal databases. *Geoinformatica*, 3(1):61–99, 1999. ISSN 1384-6175. doi: http://dx.doi.org/10.1023/A: 1009800916313.

Dimitris Achlioptas. Database-friendly random projections. *Journal of Computer System Science*, 66(4):671–687, 2003. ISSN 0022-0000. doi: http://dx.doi.org/10.1016/ S0022-0000(03)00025-4.

Rakesh Agrawal and Ramakrishnan Srikant. Fast algorithms for mining association rules in large databases. In *Proceedings of the 20th International Conference on Very Large Data Bases (VLDB)*, pages 487–499, San Francisco, CA, USA, 1994. Morgan Kaufmann Publishers Inc. ISBN 1-55860-153-8.

Rakesh Agrawal, Christos Faloutsos, and Arun Swami. Efficient similarity search in sequence databases. In *Proceedings of the 4th International Conference on Foundations of Data Organization and Algorithms (FOD)*, pages 69–84, London, UK, 1993. Springer-Verlag. ISBN 3-540-57301-1.

Ghazi Al-Naymat. Enumeration of maximal clique for mining spatial co-location patterns. In *Proceedings of the 6th ACS/IEEE International Conference on Computer Systems and Applications (AICCSA)*, pages 126–133, Doha, Qatar, 2008.

Ghazi Al-Naymat and Sanjay Chawla. Data preparation for mining complex patterns in large spatial databases. TR 576, University of Sydney, Sydney-Australia, November 2005.

Ghazi Al-Naymat and Javid Taheri. Effects of dimensionality reduction techniques on time series similarity measurements. In *Proceedings of the 6th ACS/IEEE International Conference on Computer Systems and Applications (AICCSA)*, pages 393–397, Doha, Qatar, 2008.

Ghazi Al-Naymat, Sanjay Chawla, and Joachim Gudmundsson. Dimensionality reduction for long duration and complex spatio-temporal queries. TR 600, University of Sydney, Sydney-Australia, October 2006.

Ghazi Al-Naymat, Sanjay Chawla, and Joachim Gudmundsson. Dimensionality reduction for long duration and complex spatio-temporal queries. In *Proceedings of the 2007 ACM symposium on Applied computing (ACM SAC)*, pages 393–397. ACM Press, 2007. ISBN 1-59593-480-4. doi: http://doi.acm.org/10.1145/1244002.1244095.

Ghazi Al-Naymat, Sanjay Chawla, and Joachim Gudmundsson. Random projection for mining long duration flock pattern in spatio-temporal datasets. *To Appear in GeoInformatica*, 2008a.

Ghazi Al-Naymat, Javid Taheri, and Sanjay Chawla. SparseDTW: A novel approach to speed up dynamic time warping. *Data and Knowledge Engineering*, 2008b.

Franklin Allen and Risto Karjalainen. Using genetic algorithms to find technical trading rules1. *Journal of Financial Economics*, 51(2):245–271, February 1999. available at http://ideas.repec.org/a/eee/jfinec/v51y1999i2p245-271.html.

Leonardo Angelini, F. De Carlo, C. Marangi, Mario Pellicoro, and Sebastiano Stramaglia. Clustering data by inhomogeneous chaotic map lattices. *Physical Review Letters*, 85 (3):554–557, Jul 2000. doi: 10.1103/PhysRevLett.85.554.

Sanjeev Arora and Carsten Lund. *Approximation algorithms for NP-hard problems*, chapter Hardness of approximations, pages 399–446. PWS Publishing Co., Boston, MA, USA, 1997.

Rosa I. Arriaga and Santosh Vempala. An algorithmic theory of learning: Robust concepts and random projection. In *Proceedings of the 40th Annual Symposium on Foundations of Computer Science (FOCS)*, pages 616–624, Washington, DC, USA, 1999. IEEE Computer Society. ISBN 0-7695-0409-4.

Bavani Arunasalam, Sanjay Chawla, and Pei Sun. Striking two birds with one stone: Simultaneous mining of positive and negative spatial patterns. In *Proceedings of the Fifth SIAM International Conference on Data Mining (SDM)*, pages 173–182, 2005.

Sunil Arya, David M. Mount, Nathan S. Netanyahu, Ruth Silverman, and Angela Y. Wu. An optimal algorithm for approximate nearest searching fixed dimensions. *The Journal of the ACM (JACM)*, 45(6):891–923, 1998.

Yasuo Asakura and Eiji Hato. Tracking survey for individual travel behaviour using mobile communication instruments. *Transportation Research Part C: Emerging Technologies*, 12(3-4):273–291, June 2004.

Nicolas Basalto, Roberto Bellotti, Francesco de Carlo, Paolo Facchi, and Saverio Pascazio. Clustering stock market companies via chaotic map synchronization. *Physica A: Statistical Mechanics and its Applications*, 345(1-2):196–206, 2004.

Peter N. Belhumeur, Jo ao P. Hespanha, and David J. Kriegman. Eigenfaces vs. fisherfaces: Recognition using class specific linear projection. *IEEE Transactions on Pattern Analysis and Machine Intelligence*, 19:711–720, 1997.

Richard Bellman. *Adaptive Control Processes*. Princeton University Press, 1961.

Marc Benkert, Joachim Gudmundsson, Florian Hübner, and Thomas Wolle. Reporting flock patterns. In *Proceedings of the 14th European Symposium on Algorithms (ESA)*, volume 4168/2006 of *Lecture Notes in Computer Science*, pages 660–671.

Springer Berlin / Heidelberg, 2006. doi: 10.1007/11841036. URL `http://www.`
`springerlink.com/content/87626406v8410v7v/`.

Donald J. Berndt and James Clifford. Using dynamic time warping to find patterns in time series. In *Association for the Advancement of Artificial Intelligence, Workshop on Knowledge Discovery in Databases (AAAI)*, pages 229–248, 1994.

Donald J Berndt and James Clifford. Finding patterns in time series: A dynamic programming approach. In *Proceedings of the Advances in Knowledge Discovery and Data Mining*, pages 229–248, 1996.

Ella Bingham and Heikki Mannila. Random projection in dimensionality reduction: applications to image and text data. In *Proceedings of the seventh ACM SIGKDD international conference on Knowledge discovery and data mining (KDD)*, pages 245–250. ACM Press, 2001. ISBN 1-58113-391-X. doi: http://doi.acm.org/10.1145/502512. 502546.

Christopher M. Bishop. *Pattern Recognition and Machine Learning (Information Science and Statistics)*. Springer, 2006.

Michael J. Black, Yaser Yacoob, Allan D. Jepson, and David J. Fleet. Learning parameterized models of image motion. In *Proceedings of the 1997 Conference on Computer Vision and Pattern Recognition (CVPR)*, pages 561–567, Washington, DC, USA, 1997. IEEE Computer Society. ISBN 0-8186-7822-4.

Jaron Blackburn and Eraldo Ribeiro. *Human Motion Understanding, Modeling, Capture and Animation*, volume 4814/2007 of *Lecture Notes in Computer Science*, chapter Human Motion Recognition Using Isomap and Dynamic Time Warping, pages 285–298. Springer Berlin / Heidelberg, November 2007.

John S Boreczky and Lawrence A. Rowe. Comparison of video shot boundary detection techniques. *Journal of Electronic Imaging*, 5(2):122–128, April 1996.

Kevin Buchin, Maike Buchin, and Joachim Gudmundsson. Detecting single file movement. In *Proceedings of the 16th ACM Conference on Advances in Geographic Information Systems (ACM GIS)*, pages 288–297, 2008a.

Kevin Buchin, Maike Buchin, Joachim Gudmundsson, Maarten Löffler, and Jun Luo. Detecting commuting patterns by clustering subtrajectories. In *Proceedings of the 19th International Symposium on Algorithms and Computation (ISAAC)*, volume 5369 of *Lecture Notes in Computer Science*, pages 644–655. Springer, 2008b.

Jeremy Buhler and Martin Tompa. Finding motifs using random projections. In *Proceedings of the fifth annual international conference on Computational biology (RE-COMB)*, pages 69–76, New York, NY, USA, 2001. ACM. ISBN 1-58113-353-7. doi: http://doi.acm.org/10.1145/369133.369172.

EG Caiani, A Porta, G Baselli, M Turie, S Muzzupappa, Piemzzi, C Crema, A Malliani, and S Cerutti. Warped-average template technique to track on a cycle-by-cycle basis the cardiac filling phases on left ventricular volume. *Computers in Cardiology*, 5: 73–76, 1998.

Huiping Cao, Nikos Mamoulis, and David W. Cheung. Mining frequent spatio-temporal sequential patterns. In *Proceedings of the Fifth IEEE International Conference on Data Mining (ICDM)*, pages 82–89, Washington, DC, USA, 2005. IEEE Computer Society. ISBN 0-7695-2278-5. doi: http://dx.doi.org/10.1109/ICDM.2005.95.

Longbing Cao, Chao Luo, Jiarui Ni, Dan Luo, and Chengqi Zhang. Stock data mining through fuzzy genetic algorithms. In *Proceedings of the 9th Joint Conference on Information Sciences (JCIS)*, Advances in Intelligent Systems Researc, 2006a.

Longbing Cao, Dan Luo, and Chengqi Zhang. Fuzzy genetic algorithms for pairs mining. In *Proceedings of the 9th Pacific Rim International Conference on Artificial Intelligence (PRICAI)*, volume 4099 of *Lecture Notes in Computer Science*, pages 711–720. Springer Berlin / Heidelberg, 2006b.

Paolo Capitani and Paolo Ciaccia. Warping the time on data streams. *Data and Knowledge Engineering*, 62(3):438–458, 2007. ISSN 0169-023X. doi: http://dx.doi.org/10.1016/j.datak.2006.08.012.

Chris Chatfield. *The Analysis of Time Series: An Introduction*. CRC press, sixth edition, 2004.

An-Pin Chen, Sheng-Fuu Lin, and Yi-Chang Cheng. Time registration of two image sequences by dynamic time warping. *IEEE International Conference on Networking, Sensing and Control*, 1(21–23):418 – 423, March 2004.

Shu-Heng Chen. *Genetic Algorithms and Genetic Programming in Computational Finance*. Kluwer Academic, 2002.

Jin Cheqing, Xiong Fang, Huang Joshua Zhexue, Yu Jeffrey Xu, and Zhou Aoying. Mining frequent items in spatio-temporal databases. *Lecture Notes in Computer Science*, 3129:549–558, 2004.

Benny Chor and Madhu Sudan. A geometric approach to betweenness. *SIAM Journal on Discrete Mathematics (SIDMA)*, 11(4):511–523, 1998. ISSN 0895-4801. doi: http://dx.doi.org/10.1137/S0895480195296221.

David A. Coley. *An Introduction to Genetic Algorithms for Scientists and Engineers*. World scientific publishing co, 1999.

Timothy F. Cootes, Gareth J. Edwards, and Christopher J. Taylor. Active appearance models. *IEEE Transactions on Pattern Analysis and Machine Intelligence*, 22(6):681–685, June 2001.

Steve Donoho. Early detection of insider trading in option markets. In *Proceedings of the tenth international conference on Knowledge discovery and data mining (ACM SIGKDD)*, pages 420–429, New York, NY, USA, 2004. ACM Press. ISBN 1-58113-888-1. doi: http://doi.acm.org/10.1145/1014052.1014100.

Cedric du Mouza and Philippe Rigaux. Mobility patterns. *GeoInformatica*, 9(4):297–319, December 2005.

Bertrand Dumont, Alain Boissy, C Achard, A Sibbald, and H.W. Erhard. Consistency of animal order in spontaneous group movements allows the measurement of leadership in a group of grazing heifers. *Applied Animal Behaviour Science*, 95(1–2):55–66, 2005.

Max Egenhofer. What's special about spatial?: database requirements for vehicle navigation in geographic space. In *Proceedings of the ACM SIGMOD international conference on Management of data (SIGMOD)*, pages 398–402, New York, NY, USA, 1993. ACM Press. ISBN 0-89791-592-5. doi: http://doi.acm.org/10.1145/170035.170096.

Mahmoud Mohamed Abd Ellatif. Association rules technique to diagnosis financial performance for ksa stock market companies, December 2007. URL http://ssrn.com/abstract=899023.

David Eppstein, Michael T. Goodrich, and Jonathan Z. Sun. The skip quadtree: A simple dynamic data structure for multidimensional data. In *Proceedings of ACM Symposium on Computational Geometry (SCG)*, pages 296–305, Pisa, 2005. URL http://www.citebase.org/abstract?id=oai:arXiv.org:cs/0507049.

Christos Faloutsos, Mudumbai Ranganathan, and Yannis Manolopoulos. Fast subsequence matching in time-series databases. *SIGMOD Record*, 23(2):419–429, 1994. ISSN 0163-5808. doi: http://doi.acm.org/10.1145/191843.191925.

Xiaoli Zhang Fern and Carla E. Brodley. Random projection for high dimensional data clustering: A cluster ensemble approach. In *Proceedings of the 20th International Conference on Machine Learning (ICML)*, pages 186–193, August 2003.

Dmitriy Fradkin and David Madigan. Experiments with random projections for machine learning. In *Proceedings of the ninth ACM SIGKDD international conference on Knowledge discovery and data mining (KDD)*, pages 517–522. ACM Press, 2003. ISBN 1-58113-737-0. doi: http://doi.acm.org/10.1145/956750.956812.

Andrew Frank. *Life and Motion of Socio-Economic Units*, chapter 2: Socio-Economic Units: Their Life and Motion, pages 21–34. Taylor & Francis, 2001.

Peter Frankl and Hiroshi Maehara. The johnson-lindenstrauss lemma and the sphericity of some graphs. *Journal of Combinatorial Theory Series A*, 44(3):355–362, 1987. ISSN 0097-3165.

Evan Gatev, William N. Goetzmann, and K. Geert Rouwenhorst. Pairs trading: Performance of a relative-value arbitrage rule. *Published by Oxford University Press on behalf of The Society for Financial Studies*, 19(3):797–827, February 2006.

Navin Goel, George Bebis, and Ara Nefian. Face recognition experiments with random projection. In *Proceedings of Society of Photo-Optical Instrumentation Engineers Conference Series (SPIE)*, volume 5779, pages 426–437, Bellingham, WA, 2005.

Jim Gray, Don Slutz, Alex S. Szalay, Ani R. Thakar, Jan vandenBerg, Peter Z. Kunszt, and Christopher Stoughton. Data mining the sdss skyserver database. Technical Report MSR-TR-2002-01, Microsoft Research, 2002.

Joachim Gudmundsson and Marc van Kreveld. Computing longest duration flocks in trajectory data. In *Proceedings of the 14th annual ACM international symposium on Advances in geographic information systems (GIS)*, pages 35–42. ACM Press, 2006. ISBN 1-59593-529-0. doi: http://doi.acm.org/10.1145/1183471.1183479.

Joachim Gudmundsson, Marc van Kreveld, and Bettina Speckmann. Efficient detection of motion patterns in spatio-temporal data sets. *GeoInformatica*, 11(2):195–215, 2007.

Joachim Gudmundsson, Patrick Laube, and Thomas Wolle. Movement patterns in spatio-temporal data. In Shashi Shekhar and Hui Xiong, editors, *Encyclopedia of GIS*, pages 726–732. Springer US, March 2008.

Ralf Hartmut Güting. An introduction to spatial database systems. *The VLDB Journal*, 3 (4):357–399, 1994. ISSN 1066-8888.

Ralf Hartmut Güting and Markus Schneider. *Moving Objects Databases*. Morgan Kaufmann, 2005.

Ralf Hartmut Güting, Michael H. Böhlen, Martin Erwig, Christian S. Jensen, Nikos A. Lorentzos, Markus Schneider, and Michalis Vazirgiannis. A foundation for representing and querying moving objects. *ACM Transactions on Database Systems (TODS)*, 25(1):1–42, 2000. ISSN 0362-5915. doi: http://doi.acm.org/10.1145/352958.352963.

Ralf Hartmut Güting, Michael H. Böhlen, Martin Erwig, Christian S. Jensen, Nikos Lorentzos, Enrico Nardelli, Markus Schneider, and Jose R.R. Viqueira. *Spatio-Temporal Databases: The CHOROCHRONOS Approach*, volume Volume 2520/2003, chapter 4: Spatio-temporal Models and Languages: An Approach Based on Data Types, pages 117–176. Springer Berlin / Heidelberg, 2003.

Jiawei Han and Micheline Kamber. *Data Mining: Concepts and Techniques*. Morgan Kaufmann, 2 edition, March 2006. ISBN 1-55860-901-6.

Jiawei Han, Jian Pei, and Yiwen Yin. Mining frequent patterns without candidate generation. In *Proceedings of the International Conference on Management of Data (ACM SIGMOD)*, pages 1–12. ACM Press, May 2000. ISBN 1-58113-218-2. URL citeseer.ist.psu.edu/han99mining.html.

Daniel Herlemont. Pairs trading, convergence trading, cointegration, June 2004. URL http://www.yats.com/doc/cointegration-en.pdf.

Dan Hirschberg. A linear space algorithm for computing maximal common subsequences. *Communications of the ACM*, 18(6):341–343, 1975. ISSN 0001-0782. doi: http://doi.acm.org/10.1145/360825.360861.

Harold Hotelling. Analysis of a complex of statistical variables into principal components. *Journal of Educational Psychology*, 24:417–441, 1933.

Yan Huang, Hui Xiong, Shashi Shekhar, and Jian Pei. Mining confident co-location rules without a support threshold. In *Proceedings of the ACM symposium on Applied computing (ACM SAC)*, pages 497–501, New York, NY, USA, 2003. ACM. ISBN 1-58113-624-2. doi: http://doi.acm.org/10.1145/952532.952630.

Yan Huang, Shashi Shekhar, and Hui Xiong. Discovering spatial co-location patterns from spatial data sets:a general approach. *IEEE Transaction on Knowledge and Data Engineering (TKDE)*, 16(2):1472–1485, 2004.

Piotr Indyk and Rajeev Motwani. Approximate nearest neighbors: towards removing the curse of dimensionality. In *Proceedings of the 30th ACM Symposium on Theory of Computing (STOC)*, pages 604–613. ACM Press, 1998.

Yoshiharu Ishikawa, Yuichi Tsukamoto, and Hiroyuki Kitagawa. Extracting mobility statistics from indexed spatio-temporal datasets. In *Second Workshop on Spatio-Temporal Database Management (STDBM)*, pages 9–16, 2004.

Fumitada Itakura. Minimum prediction residual principle applied to speech recognition. *IEEE Transactions on Acoustics, Speech and Signal Processing*, 23(1):67–72, 1975.

Christian S. Jensen, Dan Lin, and Beng Chin Ooi. Continuous clustering of moving objects. *IEEE Transactions on Data Engineering*, 19(9):1161–1174, 2007.

Mennis Jeremy and Liu Jun Wei. Mining association rules in spatio-temporal data: An analysis of urban socioeconomic and land cover change. *Transactions in GIS*, 9(1): 5–17, 2005. ISSN j.1467-9671.2005.00202.x.

Hoyoung Jeung, Heng Tao Shen, and Xiaofang Zhou. Mining trajectory patterns using hidden markov models. In *Proceedings of the 9th International Conference on Data Warehousing and Knowledge Discovery (DaWaK)*, volume 4654 of *Lecture Notes in Computer Science*, pages 470–480, 2007.

Hoyoung Jeung, Man Lung Yiu, Xiaofang Zhou, Christian Jensen, and Heng Tao Shen. Discovery of convoys in trajectory databases. In *Proceedings of the 34th International Conference on Very Large Data Bases (VLDB)*, pages 1068–1080. ACM Press, 2008.

William B. Johnson and Joram Lindenstrauss. Extensions of lipschitz mappings into a hilbert space. In *Conference in modern analysis and probability (New Haven, Conn., 1982)*, pages 189–206. American Mathematical Society, 1982. doi: http://doi.acm.org/ 10.1145/502512.502546.

Damon D. Judd. What's so special about spatial data?, 2005.

Panos Kalnis, Nikos Mamoulis, and Spiridon Bakiras. On discovering moving clusters in spatio-temporal data. In *Proceedings of the 9th International Symposium on Advances in Spatial and Temporal Databases (SSTD)*, pages 364–381, 2005.

Samuel Kaski. Dimensionality reduction by random mapping: Fast similarity computation for clustering. In *Proceedigns of the International Joint Conference on Neural*

Networks (IJCNN), volume 1, pages 413–418, Piscataway, NJ, 1998. IEEE Service Center. URL `citeseer.ist.psu.edu/kaski98dimensionality.html`.

Eamonn Keogh. The ucr time series data mining archive. http://www.cs.ucr.edu/~eamonn/TSDMA/index.html, Septemper 2006.

Eamonn Keogh and Michael Pazzani. Scaling up dynamic time warping for datamining applications. In *Proceedings of the sixth ACM SIGKDD international conference on Knowledge discovery and data mining (KDD)*, pages 285–289, New York, NY, USA, 2000. ACM Press. ISBN 1-58113-233-6. doi: http://doi.acm.org/10.1145/347090. 347153.

Eamonn Keogh and Chotirat Ratanamahatana. Exact indexing of dynamic time warping. *Knowledge and Information Systems (KIS)*, 7(3):358–386, 2004. ISSN 0219-1377. doi: http://dx.doi.org/10.1007/s10115-004-0154-9.

Sang-Wook Kim, Sanghyun Park, and Wesley Chu. An index-based approach for similarity search supporting time warping in large sequence databases. In *Proceedings of the 17th International Conference on Data Engineering (ICDE)*, pages 607–614, Washington, DC, USA, 2001. IEEE Computer Society. ISBN 0-7695-1001-9.

Jon M. Kleinberg. Two algorithms for nearest-neighbor search in high dimensions. In *Proceedings of the twenty-ninth annual ACM symposium on Theory of computing (STOC)*, pages 599–608, New York, NY, USA, 1997. ACM. ISBN 0-89791-888-6. doi: http://doi.acm.org/10.1145/258533.258653.

George Kollios, Stan Sclaroff, and Margrit Betke. Motion mining: Discovering spatio-temporal patterns in databases of human motion. In *Workshop on Research Issues in Data Mining and Knowledge Discovery (DMKD)*, 2001.

Krzysztof Koperski, Junas Adhikary, and Jiawei Han. Spatial data mining: Progress and challenges. In *Proceedings of the SIGMOD Workshop on Research Issues on data Mining and Knowledge Discovery (DMKD*, pages 1–10, 1996.

Manolis Koubarakis, Yannis Theodoridis, and Timos Sellis. *Spatio-Temporal Databases*, volume 2520/2003 of *Lecture Notes in Computer Science*, chapter 9: Spatio-temporal Databases in the Years Ahead, pages 345–347. Springer Berlin / Heidelberg, 2003.

Z.M. Kovacs-Vajna. A fingerprint verification system based on triangular matching and dynamic time warping. *IEEE Transactions on Pattern Analysis and Machine Intelligence*, 22(11):1266–1276, November 2000.

Boris Kovalerchuk and Evgenii Vityaev. *Data Mining in Finance: Advances in Relational and Hybrid Methods*. Kluwer Academic Publishers, 2000.

Hans-Peter Kriegel. Knowledge discovery in spatial databases. 2005.

Anukool Lakhina, Mark Crovella, and Christiphe Diot. Characterization of network-wide anomalies in traffic flows. In *Proceedings of the 4th ACM SIGCOMM conference on Internet measurement (IMC)*, pages 201–206, New York, NY, USA, 2004a. ACM. ISBN 1-58113-821-0. doi: http://doi.acm.org/10.1145/1028788.1028813.

Anukool Lakhina, Mark Crovella, and Christophe Diot. Diagnosing network-wide traffic anomalies. In *Proceedings of the 2004 conference on Applications, technologies, architectures, and protocols for computer communications (SIGCOMM)*, pages 219–230, New York, NY, USA, 2004b. ACM. ISBN 1-58113-862-8. doi: http://doi.acm.org/10.1145/1015467.1015492.

Patrick Laube and Stephan Imfeld. Analyzing relative motion within groups of trackable moving point objects. In *Proceedings of the second International Conference on Geographic Information Science (GIS)*, pages 132–144. Springer-Verlag, 2002. ISBN 3-540-44253-7.

Patrick Laube, Marc van Kreveld, and Stephan Imfeld. Finding REMO – detecting relative motion patterns in geospatial lifelines. In *Proceedings of the eleventh International Symposium on Spatial Data Handling (SDH)*, pages 201–215. Springer Berlin
Heidelberg, 2004.

Ramon Lawrence. Using neural networks to forecast stock market prices, December 1997. URL http://www.cs.uiowa.edu/~rlawrenc/research/
Papers/nn.pdf.

Ping Li, Trevor J. Hastie, and Kenneth W. Church. Very sparse random projections.
In *Proceedings of the 12th ACM SIGKDD international conference on Knowledge
discovery and data mining (KDD)*, pages 287–296, New York, NY, USA, 2006. ACM.
ISBN 1-59593-339-5. doi: http://doi.acm.org/10.1145/1150402.1150436.

Li Lin, Longbing Cao, Jiaqi Wang, and Chengqi Zhang. The applications of genetic
algorithms in stock market data mining optimisation. *Information and Communication
Technologies*, 33:9, 2004.

Li Lin, Longbing Cao, and Chengqi Zhang. Genetic algorithms for robust optimization
in financial applications. In M.H. Hamza, editor, *Computational Intelligence*, 2005.

Bertis B. Little, Walter L. Johnston, Ashley C. Lovell, Roderick M. Rejesus, and Steve A.
Steed. Collusion in the u.s. crop insurance program: applied data mining. In *Proceedings of the eighth ACM SIGKDD international conference on Knowledge discovery
and data mining (KDD)*, pages 594–598, New York, NY, USA, 2002. ACM Press.
ISBN 1-58113-567-X. doi: http://doi.acm.org/10.1145/775047.775135.

Kun Liu and Jessica Ryan. Random projection-based multiplicative data perturbation
for privacy preserving distributed data mining. *IEEE Transactions on Knowledge and
Data Engineering (TKDE)*, 18(1):92–106, 2006. ISSN 1041-4347. doi: http://dx.doi.
org/10.1109/TKDE.2006.14. Senior Member-Hillol Kargupta.

Ines Fernando Vega Lopez and Bongki Moon. Spatiotemporal aggregate computation: A survey. *IEEE Transactions on Knowledge and Data Engineering (TKDE)*, 17(2):271–286, 2005. ISSN 1041-4347. doi: http://dx.doi.org/10.1109/TKDE.2005.34. Senior Member-Richard T. Snodgrass.

Hongjun Lu, Jiawei Han, and Ling Feng. Stock movement prediction and n-dimensional inter-transaction association rules. In *ACM SIGMOD Workshop on Research Issues on Data Mining and Knowledge Discovery (SIGMOD)*, pages 121–127, Seattle,Washington, 1998.

Haynes M. and S. Churchman. Hubble's law. http://map.gsfc.nasa.gov/, March 2005. URL http://map.gsfc.nasa.gov/.

Scott Makeig, Marissa Westerfield, Jeanne Townsend, Tzyy-Ping Jung, Eric Courchesne, and Terrence Sejnowski. Functionally independent components of early event-related potentials in a visual spatial attention task. *Philosophical Transaction of The Royal Society: Bilogical Science*, 354(1387):1135–1144, 1999.

Luis Malagón-Borja and Olac Fuentes. An object detection system using image reconstruction with pca. In *Proceedings of the 2nd Canadian conference on Computer and Robot Vision (CRV)*, pages 2–8, Washington, DC, USA, 2005. IEEE Computer Society. ISBN 0-7695-2319-6. doi: http://dx.doi.org/10.1109/CRV.2005.16.

Nikos Mamoulis, Huiping Cao, George Kollios, Marios Hadjieleftheriou, Yufei Tao, and David W. Cheung. Mining, indexing, and querying historical spatiotemporal data. In *Proceedings of the tenth ACM SIGKDD international conference on Knowledge discovery and data mining (KDD)*, pages 236–245. ACM Press, 2004. ISBN 1-58113-888-1. doi: http://doi.acm.org/10.1145/1014052.1014080.

Vicent Martin and Enn Saar. *Statistics of the Galaxy Distribution*. Chapman and Hall/CRC, 2002. ISBN 1584880848.

Baback Moghaddam and Alex Sandy Pentland. Probabilistic visual learning for object detection. In *Proceedings of the Fifth International Conference on Computer Vision (ICCV)*, pages 786–793, Washington, DC, USA, 1995. IEEE Computer Society. ISBN 0-8186-7042-8.

Yasuhiko Morimoto. Mining frequent neighboring class sets in spatial databases. In *Proceedings of the Seventh ACM SIGKDD international conference on Knowledge discovery and data mining (KDD)*, pages 353–358. ACM Press, 2001.

Mario E. Munich and Pietro Perona. Continuous dynamic time warping for translation-invariant curve alignment with applications to signature verification. In *Proceedings of the 8th IEEE International Conference on Computer Vision*, pages 108–115, 1999.

Rob Munro, Sanjay Chawla, and Pei Sun. Complex spatial relationships. In *Proceedings of the 3rd IEEE International Conference on Data Mining (ICDM)*, pages 227–234. IEEE Computer Society, 2003. ISBN 0-7695-1978-4.

Hiroshi Murase and Shree K. Nayar. Visual learning and recognition of 3-d objects from appearance. *International Journal of Computer Vision*, 14(1):5–24, 1995. ISSN 0920-5691. doi: http://dx.doi.org/10.1007/BF01421486.

Cory Myers, Lawrence R. Rabiner, and Aaron E. Rosenberg. Performance tradeoffs in dynamic time warping algorithms for isolated word recognition. *IEEE Transactions on Acoustics, Speech and Signal Processing*, 28(6):623–635, December 1980.

Keith Nesbitt and Stephen Barrass. Finding trading patterns in stock market data. *IEEE Computer Graphics and Applications*, 24(5):45–55, 2004. ISSN 0272-1716. doi: http://dx.doi.org/10.1109/MCG.2004.28.

Raymond Ng. *Geographic Data Mining and Knowledge Discovery*, chapter 9: Detecting Outliers form Large Datasets, pages 218–235. Taylor & Francis, 2001.

Nuria M. Oliver, Barbara Rosario, and Alex P. Pentland. A bayesian computer vision system for modeling human interactions. *IEEE Transactions on Pattern Analysis and Machine Intelligence*, 22(8):831–843, 2000. ISSN 0162-8828. doi: http://dx.doi.org/10.1109/34.868684.

Hyun Kyoo Park, Jin Hyun Son, and Myoung-Ho Kim. An efficient spatiotemporal indexing method for moving objects in mobile communication environments. In *Proceedings of the 4th International Conference on Mobile Data Management (MDM)*, pages 78–91, London, UK, 2003. Springer-Verlag. ISBN 3-540-00393-2.

Dieter Pfoser, Christian S. Jensen, and Yannis Theodoridis. Novel approaches in query processing for moving object trajectories. In *Proceedings of the 26th International Conference on Very Large Data Bases (VLDB)*, pages 395–406, San Francisco, CA, USA, 2000. Morgan Kaufmann Publishers Inc. ISBN 1-55860-715-3.

Gregory Piatetsky-Shapiro and William J. Frawley. *Knowledge Discovery in Databases*. AAAI/MIT Press, 1991. ISBN 0-262-62080-4.

Yunyao Qu, Changzhou Wang, and X. Sean Wang. Supporting fast search in time series for movement patterns in multiple scales. In *Proceedings of the seventh international conference on Information and knowledge management (CIKM)*, pages 251–258, New York, NY, USA, 1998. ACM Press. ISBN 1-58113-061-9. doi: http://doi.acm.org/10.1145/288627.288664.

Lawrence Rabiner and Biing-Hwang Juang. *Fundamentals of speech recognition*. Prentice Hall Signal Processing Series, Upper Saddle River, NJ, USA, 1993. ISBN 0-13-015157-2.

Davood Rafiei and Alberto Mendelzon. Efficient retrieval of similar time sequences using dft. In *Proceedings of the In Intertional Conference on Foundations of Data Organization and Algorithms (FODO)*, pages 249–257, 1998.

Haakon Ringberg, Augustin Soule, Jennifer Rexford, and Christophe Diot. Sensitivity of pca for traffic anomaly detection. In *Proceedings of the 2007 ACM SIGMETRICS international conference on Measurement and modeling of computer systems (SIGMETRICS)*, pages 109–120, New York, NY, USA, 2007. ACM. ISBN 978-1-59593-639-4. doi: http://doi.acm.org/10.1145/1254882.1254895.

Ali A. Roshannejad and W. Kainz. Handling identities in spatio- temporal databases. In *Proceedings of the Twelfth International Symposium on Computer- Assisted Cartography*, pages 119–126, Charlotte, North Carolina, 1995.

Davide Roverso. Multivariate temporal classification by windowed wavelet decomposition and recurrent neural networks. In *Proceedings of the 3rd ANS International Topical Meeting on Nuclear Plant Instrumentation, Control and Human-Machine Interface Technologies (NPIC and HMIT)*, 2000.

Hiroaki Sakoe and Seibi Chiba. Dynamic programming algorithm optimization for spoken word recognition. *IEEE Transactions on Acoustics, Speech and Signal Processing*, 26(1):43–49, 1978.

Yasushi Sakurai, Masatoshi Yoshikawa, and Christos Faloutsos. FTW: Fast similarity search under the time warping distance. In *Proceedings of the twenty-fourth ACM SIGMOD-SIGACT-SIGART symposium on Principles of database systems (PODS)*, pages 326–337, New York, NY, USA, 2005. ACM. ISBN 1-59593-062-0. doi: http://doi.acm.org/10.1145/1065167.1065210.

Stan Salvador and Philip Chan. Toward accurate dynamic time warping in linear time and space. *Intelligent Data Analysis*, 11(5):561–580, 2007.

Hanan Samet. *Foundations of Multidimensional and Metric Data Structures*. Morgan Kaufmann, 2006.

David Sankoff and Joseph Kruskal. *Time Warps, String Edits, and Macromolecules: The Theory and Practice of Sequence Comparison*. Addsion Wisely Press,, 1999.

Mohamad H. Saraee and Babis Theodoulidis. Knowledge discovery in temporal databases. In *IEE Colloquium on Digest No. 1995/021(A)*, pages 1–4. February 1995.

Matthew Schmill, Tim Oates, and Paul Cohen. Learned models for continuous planning. In *The Seventh International Workshop on Artificial Intelligence and Statistics (AISTATS)*, pages 278–282, 1999.

Shashi Shekhar and Yan Huang. Discovering spatial co-location patterns: A summary of results. *Lecture Notes in Computer Science*, 2121:236–256, 2001. URL citeseer. ifi.unizh.ch/shekhar01discovering.html.

Takeshi Shirabe. Correlation analysis of discrete motions. In *Proceedings of the 4th International Conference on Geographic Information Science (GIScience)*, pages 370–382, 2006.

David Spergel, Michael Bolte, and Wendy Freedman. The age of the universe. *Proceedings of the National Academy of Science*, 94(13):6579–6584, Jun 24 1997.

Barry Storey and Robert Holtom. The use of historic gps data in transport and traffic monitoring. *Traffic Engineering and Control*, 44(10):376–379, November 2003.

Sloan Digital Sky Survey. Sdss - sloan digital sky survey. retrieved august 5, 2005 from http://cas.sdss.org/dr3/en/help/download/, 2005.

Javid Taheri and Albert Y. Zomaya. Realistic simulations for studying mobility management problems. *International Journal of Wireless and Mobile Computing*, 1(8), 2005.

Pang-Ning Tan, Michael Steinbach, and Vipin Kumar. *Introduction to Data Mining*. Pearson Addison-Wesley, 2006.

Chunqiang Tang, Sandhya Dwarkadas, and Zhichen Xu. On scaling latent semantic indexing for large peer-to-peer systems. In *Proceedings of the 27th annual international ACM SIGIR conference on Research and development in information retrieval (SIGIR)*, pages 112–121, New York, NY, USA, 2004. ACM. ISBN 1-58113-881-4. doi: http://doi.acm.org/10.1145/1008992.1009014.

Yufei Tao and Dimitris Papadias. Mv3r-tree: A spatio-temporal access method for timestamp and interval queries. In *Proceedings of the 27th International Conference on Very Large Data Bases (VLDB)*, pages 431–440, San Francisco, CA, USA, 2001. Morgan Kaufmann Publishers Inc. ISBN 1-55860-804-4.

Yufei Tao, George Kollios, Jeffrey Considine, Feifei Li, and Dimitris Papadias. Spatio-temporal aggregation using sketches. In *Proceedings of the 20th International Conference on Data Engineering (ICDE)*, pages 214–225, Washington, DC, USA, 2004. IEEE Computer Society. ISBN 0-7695-2065-0.

Charles C. Tappert and Subrata K. Das. Memory and time improvements in a dynamic programming algorithm for matching speech patterns. *IEEE Transactions on Acoustics, Speech and Signal Processing*, 26(6):583–586, December 1978.

Bhavani Thuraisingham. *Data Mining: Technologies, Techniques, Tools and Trends*. CRC Press, 1998. ISBN 0-8493-1815-7.

Yann Tremblay, Scott A. Shaffer, Shannon L. Fowler, Carey E. Kuhn, Birgitte I. McDonald, Michael J. Weiseand Charle-Andr Bostand Henri Weimerskirch, Daniel E. Crocker, Michael E. Goebel, and Daniel P. Costa. Interpolation of animal tracking data in a fluid environment. *Journal of Experimental Biology*, 209:128–140, 2006.

Ilias Tsoukatos and Dimitrios Gunopulos. Efficient mining of spatiotemporal patterns. In *Proceedings of the 7th International Symposium on Advances in Spatial and Temporal*

Databases (SSTD), pages 425–442, London, UK, 2001. Springer-Verlag. ISBN 3-540-42301-X.

Thierry Urruty, Chabane Djeraba, and Dan A. Simovici. Clustering by random projections. In *Advances in Data Mining. Theoretical Aspects and Applications*, volume 4597/2007, pages 107–119. 2007.

Ronald Vanderlinden. Sunspot data. http://sidc.oma.be/html/sunspot.html, May 2008. URL http://sidc.oma.be/html/sunspot.html.

Santosh Vempala. Random projection: A new approach to vlsi layout. In *Proceedings of the 39th Annual Symposium on Foundations of Computer Science (FOCS)*, pages 389–396, Washington, DC, USA, 1998. IEEE Computer Society. ISBN 0-8186-9172-7.

Florian Verhein and Ghazi Al-Naymat. Fast mining of complex spatial co-location patterns using glimit. In *The 2007 International Workshop on Spatial and Spatio-temporal Data Mining (SSTDM) in cooperation with The 2007 IEEE International Conference on Data Mining (ICDM)*, pages 679–684, Los Alamitos, CA, USA, 2007. IEEE Computer Society. ISBN 0-7695-3033-8.

Florian Verhein and Sanjay Chawla. Geometrically inspired itemset mining. In *Proceedings of the International Conference on Data Mining (ICDM)*, pages 655–666. IEEE Computer Society, 2006. URL http://doi.ieeecomputersociety.org/10.1109/ICDM.2006.75.

Florian Verhein and Sanjay Chawla. Mining spatio-temporal patterns in object mobility databases. *Data Mining and Knowledge Discovery (DMKD)*, 16(1):5–38, February 2008.

Granapathy Vidyamurthy. *Pairs Trading Quantitative Methods and Analysis*. Wiley, 2004.

Yida Wang, Ee-Peng Lim, and San-Yih Hwang. On mining group patterns of mobile users. In *Proceedings of the 14th International Conference on Database and eXpert Systems Applications (DEXA)*, pages 287–296, Prague, September 2003.

ANDREAS S. WEIGEND. Data mining in finance: Report from the post-nncm-96 workshop on teaching computer intensive methods for financial modeling and data analysis. In *Proceedings of the Fourth International Conference on Neural Networks in the Capital Markets (NNCM)*, pages 399–412, 1996.

Halbert White. Economic prediction using neural networks: the case of IBM daily stock returns. In *IEEE International Conference on Neural Networks (ICNN)*, volume 2, pages 451–458, 1988.

Ian H. Witten and Eibe Frank. *Data Mining: Practical Machine Learning Tools and Techniques with Java Implementations*. The Morgan Kaufmann, 2000.

Jean Wolf, Stefan Schoenfelder, Marcelo Gurgel Simas de Oliveira, and Kay W Axhausen. Eighty weeks of global positioning system traces: Approaches to enriching trip information. *Transportation Research Record: Journal of the Transportation Research Board*, 1870:46–54, January 2004.

Ouri Wolfson and Eduardo Mena. *Spatial Databases: Technologies, Techniques and Trends*, chapter VIII Applications of Moving Objects Databases, pages 186–203. Idea Group Publishing, 2004.

Ouri Wolfson, Bo Xu, Sam Chamberlain, and Liqin Jiang. Moving objects databases: Issues and solutions. In *Proceedings of the Tenth International Conference on Scientific and Statistical Database Management (SSDBM)*, pages 111–122, 1998.

X. Xu, J. Han, and W. Lu. Rt-tree: an improved r-tree index structure for spatiotemporal databases. In *In Proceedings of the 4th International Symposium on Spatial Data Handling*, volume 2, pages 1040–1049, Zurich, Switzerland, July 1990.

Byoung-Kee Yi, Hosagrahar V. Jagadish, and Christos Faloutsos. Efficient retrieval of similar time sequences under time warping. In *Proceedings of the Fourteenth International Conference on Data Engineering (ICDE)*, pages 201–208, Washington, DC, USA, 1998. IEEE Computer Society. ISBN 0-8186-8289-2.

Lotfi A. Zadeh. Fuzzy sets. *Information and Control*, 8:338–353, 1965.

Xin Zhang, Nikos Mamoulis, David W. Cheung, and Yutao Shou. Fast mining of spatial collocations. In *Proceedings of the tenth ACM SIGKDD international conference on Knowledge discovery and data mining (KDD)*, pages 384–393. ACM Press, 2004. ISBN 1-58113-888-9.

Zhongfei (Mark) Zhang, John J. Salerno, and Philip S. Yu. Applying data mining in investigating money laundering crimes. In *Proceedings of the ninth international conference on Knowledge discovery and data mining (ACM SIGKDD)*, pages 747–752, New York, NY, USA, 2003. ACM Press. ISBN 1-58113-737-0. doi: http://doi.acm.org/10.1145/956750.956851.